HOW TO RESTORE FARMALL TRACTORS

By Tharran E. Gaines

Quarto.com

First Published in 2005 by Voyageur Press, an imprint of MBI Publishing Company.

New Edition published in 2021 by Motorbooks, an imprint of The Quarto Group,
100 Cummings Center, Suite 265-D, Beverly, MA 01915, USA.
T (978) 282-9590 F (978) 283-2742

Motorbooks titles are also available at discount for retail, wholesale, promotional, and bulk purchase. For details, contact the Special Sales Manager by email at specialsales@quarto.com or by mail at The Quarto Group, Attn: Special Sales Manager, 100 Cummings Center, Suite 265-D, Beverly, MA 01915, USA.

ISBN: 978-0-7603-6896-1

Digital edition published in 2021
eISBN: 978-0-7603-6897-8

Library of Congress Cataloging-in-Publication Data

Names: Gaines, Tharran E., 1950- author.
Title: How to restore Farmall tractors : the ultimate do-it-yourself guide to rebuilding and restoring / by Tharran E Gaines.
Other titles: How to restore classic Farmall tractors
Description: Beverly : Motorbooks, 2021. | Series: Motorbooks workshop | First published in 2005 as How to restore classic Farmall tractors by Voyageur Press, an imprint of MBI Publishing Company.
Identifiers: LCCN 2020023564 (print) | LCCN 2020023565 (ebook) | ISBN 9780760368961 (paperback) | ISBN 9780760368978 (ebook)
Subjects: LCSH: IHC tractors--Conservation and restoration.
Classification: LCC TL233.6.I38 G38 2021 (print) | LCC TL233.6.I38 (ebook) | DDC 629.28/752--dc23
LC record available at https://lccn.loc.gov/2020023564
LC ebook record available at https://lccn.loc.gov/2020023565

Designed by: Maria Friedrich

On the front cover: Restored Farmall B.
(Ralph W. Sanders photo/Motorbooks Archive)

On the title page: Restored Farmall AV

Opposite: A well-used Farmall is refueled for a return to the field. (Library of Congress)

Contents

A field of red: Restored and original Farmalls grace a tractor show lot.

A restored Farmall O-12, the rare cowled version of the little F-12 designed for grove and orchard use. (Photograph by Hans Halberstadt)

Acknowledgments

Having written five different books on tractor restoration, I'm often asked if I restore tractors myself or how many tractors I have. So I'll be honest with you right up front. The only classic farm tractors I own are three vintage garden tractors and several tractors in 1/16th scale.

What I do have, besides numerous models, is nearly 40 years of experience writing about farm equipment—including a couple years as editor of the Case IH *Farm Forum* magazine—and over 10 years of experience as a technical writer. I've written owner's manuals, repair manuals, and assembly instructions for companies such as Sundstrand, Hesston, Winnebago, and Kinze. My job was to glean information from a variety of sources in marketing, engineering, product service, and the test lab and turn that into text that could be used by customers to assemble, service, or repair the machines they had purchased from my employer.

I have approached tractor restoration in much the same way. I figure the people I have talked to and photographed while preparing this manuscript have forgotten more than I could ever learn about tractor restoration. So, in reality, this book is their story, not mine.

With that in mind, I want to express my appreciation to a number of people—starting with my wife, Barb, who has not only blessed me with her encouragement and patience but has spent several hours proofreading the copy and the photo captions for typographical and grammatical errors. She even accompanied me on several visits to Farmall enthusiasts.

I also want to thank all of the Farmall and IH enthusiasts and tractor restorers who have devoted their time and knowledge to this project. Without their help, I literally would not have been able to produce this book. Unfortunately, several of them have passed away since the first printing of this book. However, since few people possess as much knowledge about Farmall tractors as they did, you will still see their comments and photos in the following pages.

Among the people who either assisted with this book or contributed their wisdom are the late Walter Bieri, from Avenue City, Missouri; Paul Cummings, from Amsterdam, Missouri; Francis Gorden, from Agra, Kansas; and Gary and Mark Manville and their late father, Larry Manville, from Dearborn, Missouri. Each of these men provided a wealth of information and went above and beyond in helping me get the necessary photos to support each chapter.

Although he bled green instead of red, I also want to thank the family of the late Estel Theis, who was an antique tractor enthusiast from Avenue City, Missouri, for his assistance, particularly with photos of painting, body work, and decals.

Other Farmall and antique tractor enthusiasts who assisted with this book are the late Rex Miller, a retired farmer from Avenue City, Missouri., who repaired tractor magnetos in his spare time; Jim Seward, a tractor restorer from Wellman, Iowa; John Hunter with Maple-Hunter Decals; Dave Henderson, a tractor repair and restoration specialist from Colo, Iowa; T. W. Cook, a Farmall collector from Georgetown, Texas; Lee Grady, former McCormick-IHC collection archivist for the Wisconsin Historical Society; and the late Bob Devling, a classic tractor collector from Elwood, Kansas.

Two more people of tremendous assistance were Chris and Kim Pratt, who write and edit the *Yesterday's Tractors* online magazine found at www.ytmag.com. Due to my limited knowledge on some restoration subjects, I drew heavily on information available on their website, which I would urge you to check out for yourself. Not only is it an excellent source of tips and advice on restoration, but they also offer a variety of parts, kits, and manuals.

Others who assisted with photos, answers, or resources include Hermie Bentrup, a paint technician in St. Joseph, Missouri; Paul Jensen, owner and CEO of Jensales in Manchester, Minnesota; the Case IH parts staff at Derr Equipment in Savannah, Missouri; and Eric Wise, with Omar Associates in Avon, Ohio.

Chapter 1

Farmall History

Due to the overwhelming popularity of Farmall tractors during their fifty-year reign, there are few of us who didn't either grow up with a red tractor or have an uncle, aunt, grandparent, or cousin who owned one. It is those fond memories that lead most people to Farmall, International Harvester, and McCormick-Deering tractors as a restoration project. Of course, there are a few who simply like the bright red color of the classic Farmall tractors and want to add one to their collection.

Choosing the appropriate Farmall or McCormick-Deering model to restore, however, can be a lot more difficult than simply deciding on the brand. During the fifty years that International Harvester built tractors adorned with the Farmall name, the company introduced more than seventy different models, depending on which ones you count as unique model designations. And that doesn't even count all the McCormick-Deering and International models that don't carry the Farmall nameplate.

Becoming knowledgeable about some of the history of the company and the Farmall brands certainly isn't necessary for tractor restoration. But it may help you with your decision on which model you want to restore. Knowing that certain tractors in the line carried the same engine or transmission may even help you find parts. As an example, did you know that the A, B, and C all use the same engine, or that the 560, much to its demise, used essentially the same rear end as the late-model M?

The Two Shall Become One

Like a number of tractor manufacturers in the 1900s, International Harvester was formed as a merger of two major farm equipment companies that were forced to become partners in order for both to survive. It happened to Holt and Best when they reluctantly joined forces to form Caterpillar in 1925; it occurred again in 1953 when Massey-Harris and Harry Ferguson, Inc. came together as Massey-Ferguson.

The merger of the McCormick Harvester Company and the Deering Company in 1902 was, at first, no less welcomed by company executives and dealers. In effect, the two merged companies continued to operate much as they had before, to the point of developing two different tractor lines—the Titan and Mogul. Most of those early tractors, however, were designed for plowing and belt work, leaving jobs like cultivation, planting, and mowing hay to a team of horses.

In an effort to mechanize those tasks, too, the two merged companies began looking at small tractors by the 1910s. The Mogul line came out with its answers in the form of the 8-16 and 12-25, while the Titan line was introduced with the 10-20 and 12-25. It wasn't until 1917 that the Mogul and Titan tractor engineers were pulled together in one department with a single goal. One factor that prompted the merger was the new Fordson tractor. Introduced the same year as IH's engineering consolidation, it was small, affordable, and proving to be serious competition for all tractor manufacturers, including International Harvester.

Another influence was the United States Justice Department, which supposedly ordered International Harvester to consolidate its dealerships, as the company was breaking anti-trust laws by owning so many Titan and Mogul dealerships in overlapping territories. The company was basically ordered to reduce its sales force to one dealer in each territory, selling one brand of tractor—International.

As a result of their combined efforts, the company introduced the International 15-30 in 1921, which was as up to date as any tractor of its time. Like the International 8-16 introduced five years earlier, the 15-30 had a four-cylinder engine. However, the engine featured replaceable cylinder liners and overhead valves. The transmission even boasted a gear drive in an oil bath. As a result, the 15-30 could easily pull a three-bottom plow. After another two years, International Harvester introduced a two-plow Model 10-20. Both International models helped establish the company and unify its dealership network in preparation for big things to come.

A 1919 International Harvester Titan 10-20, owned by Ira Matheney of Modesto, California. This tractor was built shortly after the Titan and Mogul engineering groups were merged into one department. (Photograph by Hans Halberstadt)

The McCormick-Deering 10-20 was the backbone of the IH tractor line until the birth of the Farmall. This 1931 model is owned by Mike and Juanita Huss of Sandwich, Illinois.

The First Farmall

As successful as farm tractors had become by the 1920s, there was still a desperate need on the farm for a tractor that could do it all. Every tractor to date was either heavy and powerful for belt work and for pulling large implements, or it was light and maneuverable for cultivating row crops.

Recognizing the demand for a machine that could fill both roles, International Harvester began working on a "motor cultivator" as early as 1915. Unfortunately, it didn't work well and proved too expensive to build.

By 1923–1924, however, the efforts of International Harvester engineers paid off in the form of a new tractor appropriately called the Farmall. As the name implied, it could do all that a farmer wanted it to do, from powering a thresher or sheller with its 20-horsepower belt pulley to pulling a plow with 13 horsepower at the drawbar. It even had a power takeoff to run a binder or sickle mower. Unlike other tractors of its time, the Farmall also featured wide rear-wheel spacing, high ground clearance, and a narrow front end, which allowed it to straddle two rows while the front wheels ran between rows. This new configuration, which came to be known as the tricycle or row-crop design, gave farmers the ability to cultivate two rows of crops in a single pass with a machine that could perform a wide variety of other jobs on the farm.

In short order, the Farmall changed not only the future of International Harvester, but the future of the entire tractor industry worldwide.

"Studio" portrait of the first production Farmall of 1924. Note that this pioneering machine didn't even have a decal yet; instead, its name was painted on with a stencil.

First introduced in 1924, the Farmall changed the face of farming forever, as its row-crop design influenced every other major tractor company in North America. The Regular didn't go into full production, though, until 1926.

Farmall Expands the Line

Given the success and popularity of the Farmall, it didn't take long for other manufacturers to develop their own narrow-front, row-crop tractors, including John Deere, which introduced its GP (General Purpose) model in 1929. Some other makers' models were even bigger and more powerful than the Farmall. This, of course, prompted International Harvester to develop other Farmall models, as well.

The first, which went into production in late 1931, was the F-30. Equipped with a bigger engine and an improved design, it produced 30 horsepower at the PTO/belt pulley and 20 horsepower on the drawbar. It also was about three-quarters of a ton heavier than the Regular—which became the unofficial name for all Farmall tractors built before the F Series.

International Harvester didn't stop there, however. Just a few months later, the company introduced the F-20—an improved version of the Farmall Regular. Although it used the same 220-cubic-inch engine, improvements in the head and piston design provided a greater power level of 23.11 hp on the PTO/belt and 15.38 hp on the drawbar. The F-20 also came with a four-speed transmission, in place of the three-speed, and new options, including a wide front end, narrow rear tread and, eventually, rubber tires.

In the meantime, International Harvester was discovering that the F-20 and F-30 were still too large for some farmers—particularly those who still farmed with horses. The answer came in 1932 in the form of the McCormick-Deering Farmall F-12. At around $525, it cost about a third less than the F-20 and provided enough power for a one-bottom plow, a two-row planter or cultivator, or a sickle-bar mower. In fact, a full line of "quick-detachable" implements could be "on and off in a jiffy," as Farmall advertisements promised. Available with either a gasoline and distillate fuel system, the F-12 produced a maximum of 14.59 hp on the belt/PTO and 11.81 on the drawbar.

Unlike the previous Farmall models, though, the F-12 featured a unique stub-frame/transmission design and larger-diameter wheels, which provided the desired crop clearance without the need for bull-gear drop boxes at the ends of the axle.

Finally, in 1938, the F-12 was replaced by the F-14. In many respects, it was the same tractor yet with a little more power and a raised steering wheel, which provided more operator comfort. Due to a faster governor setting, the same 113-cubic-inch engine used in the F-12 now generated 17 maximum belt horsepower and 13.24 on the drawbar. Production of the F-14 only lasted two years, though. By 1939, the F-14, along with the unchanged F-20 and F-30, would be replaced by a new generation of Farmalls.

While the Regular was most renowned for its narrow front end, the factory supposedly built a few models with a wide front axle. This model, owned by Dave Henderson of Colo, Iowa, was reportedly modified in California for vegetable farming.

When the Farmall made its debut in 1924, it was designed to replace the horse on small farms, and the machine was truly capable of "farming all," as its name implied. Not only could it handle draft and belt duties, it also boasted a rear power takeoff (PTO) for running harvesters and the crop clearance necessary for cultivating. This F-20 of 1932–1939 is owned by Ira Matheney of Modesto, California. (Photograph by Hans Halberstadt)

Introduced in 1932, the Farmall F-12 cost about a third less than the F-20 and provided enough power for a one-bottom plow, which made it ideal for the farmer who needed a smaller tractor.

In 1938, International Harvester replaced the F-12 with the F-14. In addition to having a little more power, the F-14 had a raised steering wheel, providing more operator comfort.

A New Look

One of the biggest changes to affect all farm tractors occurred in the late 1930s when tractor manufacturers followed the lead of the automobile industry and began adding style to their products. It started in 1935, when Oliver Hart-Parr introduced the Oliver 70 with its sleek, streamlined design that included louvered side panels and an enclosed grille. Within two years, other sleek designs were showing up, including the heavily chromed Graham-Bradley and the Minneapolis-Moline UDLX with its futuristic cab. IH-rival John Deere even hired Henry Dreyfuss Associates, a Madison Avenue design firm well known in the industrial design world, to help style its popular A and B models.

Not to be outdone, International Harvester brought in famous industrial designer Raymond Loewy. As the designer who would later develop the famous Studebaker car designs, Loewy was commissioned to redesign the entire line of International Harvester products, including the Farmall tractors, the Fairway models, and even the company logo.

The first fruit of his efforts was the new TD-18 crawler tractor, introduced in 1938. Within a matter of months, a new line of wheeled tractors followed, starting with the A, which replaced the F-14. Right behind it was the H, which replaced the F-20 and the M, which replaced the F-30. Like the TD-18, the A, H, and M featured bright red paint, a contoured sheet metal hood that enclosed the fuel tank and steering bolster, and a new grille that enclosed the radiator. The new tractors also boasted new safety features, like standard-equipment fenders and controls that were easier to reach.

Perhaps the most unique of Loewy's design, however, was the Model A. It used the same engine as the F-12 and F-14, but that's where the similarity ended. Unlike its predecessors, the A introduced Farmall's new Culti-Vision design that offset the engine and driveline to the left, while the operator sat on the right. This gave the driver an unobstructed view straight ahead, includ-

The IH crawler tractors were among the first to feature the new styling developed by Raymond Loewy.

ing the ground beneath the tractor, where a cultivator was mounted. To attain enough ground clearance for cultivating truck garden crops, the rear axles utilized geared drop boxes, while the wide front axle rode on extended kingpins.

Meanwhile, the new Model M and H tractors were both offered in high-crop cane tractor versions and featured a choice of gasoline- or kerosene-fueled engines. Both models also came with a choice of front axles, including the dual tricycle front or an adjustable wide front end; and featured a five-speed transmission for more speed choices in and out of the field. On the other hand, the M was the first Farmall to introduce a diesel engine in 1941. Branded as the MD, it generated 35.02 horsepower on about a third less fuel at full power than the gasoline model.

Farmall wasn't out of letters yet, though. In September 1939, International Harvester introduced the B, which was basically the same tractor as the A, but with a different configuration. On the B, the engine and drivetrain were centered, as in a conventional tractor, but the operator's platform was still positioned offset to the right, thanks to a long axle on both sides of the unit. In addition, the B was only available with a dual tricycle or single-wheel front end.

Finally, in 1948, IH introduced the Model C. In addition to replacing the B, it introduced a raised operator's platform that put the driver above the tractor instead of off to the side.

A happy farm family poses alongside its pride and joy—a hard-working Farmall H.

These Farmall M and H models owned by Doug Stropes, from Joliet, Illinois, clearly illustrate the difference in size between these two popular models.

Rated at 22.45 drawbar horsepower, the Farmall H was a popular mid-sized model from 1939 to 1953. Ironically, steel wheels were still an option much of that time.

Introduced in 1939, the Model A was the first Farmall to feature Culti-Vision. This unique design positioned the engine offset to the left while the operator's seat was offset to the right. Since the A was only available with a wide front axle, this model was obviously modified to its single front wheel configuration.

While the Farmall B differed in appearance from the A, they were basically the same tractor. The B, however, featured a dual tricycle or single-wheel front end and a powertrain that was centered from front to rear.

The Super Series

By the late 1940s, customers were starting to demand more power and features and IH was answering the call with a new Super Series, starting with the Super A in 1947. In addition to more horsepower, thanks to a faster governor setting, it featured Farmall's new Touch Control hydraulic system, which used dual lift cylinders to independently raise and lower left and right side-mounted implements. Other standard amenities on the Super A were an electric starter and lights and adjustable wheel-tread spacing.

Just three years after introducing the Model C, IH replaced it with the Super C. Available with either a narrow or wide front end, the Super C had 15 percent more power than its predecessor, thanks to a larger-bore engine. Plus, it now featured Touch-Control hydraulics as standard equipment and disc brakes.

International Harvester followed much the same pattern in 1953 with the Super H. This time, though, they increased the engine displacement by approximately 12 cubic inches, giving the tractor 30 additional horsepower. The company also added a starter and generator, electric lights, hydraulic lift, pressurized radiator, PTO, and disc brakes.

About the same time, IH also introduced the Super M. It, too, received a larger engine for about 22 percent more power, as well as standard hydraulics, PTO, and electrical system. However, the Super M had a few more tricks. For one thing, it became the first model to offer an LPG version. It was also the first Farmall to feature a Torque Amplifier option. Introduced in 1954, the Super M-TA used a clutch-controlled planetary auxiliary gearbox to provide a 1.482:1 ratio increase. In effect, it was like a modern power-shift, giving the M-TA ten speeds forward and two in reverse.

The Super A was an improved version of the A. Both models shared a number of components with the Farmall C, including the 113-cubic-inch engine.

The Super M was IH's attempt to build a more powerful model without making a lot of changes to the original version, the Farmall M. This Super M is owned by Walter Bieri of Savannah, Missouri.

The Cub

While most of the line was getting bigger, there was a new end of the spectrum that was getting smaller. In 1947, when the first Super Series models were being released, IH made an even bigger splash with its new Cub. Sometimes referred to as the "Baby Farmall," the Cub was designed for vegetable farmers, tobacco growers, and other producers who farmed fewer than 40 acres.

Similar in configuration to the Farmall A, the Cub featured a four-cylinder L-head engine that produced just under 10 horsepower on the belt and a little less than nine horsepower on the drawbar via a three-speed transmission. Unlike its larger siblings, though, the Cub has the distinction of being the only Farmall that didn't use a sleeved engine. Instead, the cylinder bores were part of the block.

Altogether, the Cub was manufactured for twelve years, at which point it was replaced by the International Cub and Cub Low-Boy. Those models continued the popular line until 1975.

Numbers Replace Letters

By the early 1950s, International Harvester was seeing the need for even more changes in an effort to compete with growing rival John Deere. The first line of defense was a new Numbered Series, beginning with the Farmall 200 released in 1954. As a replacement for the Super C, it featured the same engine, but had about 10 percent more power. It also introduced a new IH option called Hydra-Creeper. Essentially, it was a hydrostatic drive that permitted the tractor to operate at speeds down to 1/4 mile per hour.

Like the other Numbered Series that followed, the 200 also sported a new grille that featured two vertical slots and more prominent horizontal slots. Another prominent feature was stainless steel numbers and letters in place of decals.

In 1955, International Harvester began adding the rest of the Numbered Series, which included the 100 as a replacement for the A, the 300 as a replacement for the Super H, and the 400, which replaced the Super M. The designation on the Cub didn't change, but it, too, received the new look and raised metal insignia.

In the meantime, the new 300 featured a Torque Amplifier for the first time, as well as a live PTO and the Fast-Hitch two-point hydraulic lift. The 300 and 400 also boasted a

The famous Cub was the next in a line of Farmall specialty tractors designed for vegetable producers and small-acreage farmers. This unique model, which has been modified with a narrow front and a narrow rear axle, is owned by Larry Manville of Dearborn, Missouri.

sleeker, more stylish hood that really turned heads with its deep red paint and shiny, raised lettering. The 400 also came with a choice of gas, LPG, or diesel engines.

International Harvester began to make changes even faster through the 1950s—to the point that by 1956, the company had already replaced the one-year-old 300 with the 350. It was nearly the same tractor, but it featured an engine displacement increase from 169 to 175 cubic inches on the 350 gas and LPG versions. It wasn't hard to spot the difference in the models, either. The new 350 featured a cream-colored grille and a cream-colored panel on each side of the hood that served as a background for the Farmall nameplate.

Also new in 1956 was the 450, which replaced the 400. Like the 350, it also got a boost in displacement, putting it at almost 50 horsepower. Other changes included a larger fuel tank and the new color scheme.

Rounding out the updates were the 130 and 230, which replaced the 100 and 200. Both feature a cream-colored grille only and benefited in terms of power from an increase in compression ratio.

Replacing the Farmall H, the 300 was one of an updated line of models that included Torque Amplifier, a live PTO and Fast-Hitch two-point hydraulic lift as standard equipment. This high-crop model is owned by Lyle Dumont of Sigourney, Iowa.

Equipped with a 281-cubic-inch gas, diesel, or LPG engine, the 450 put Farmall in the 50-horsepower class. The 450 also had a larger fuel tank than the previous 400 model. This 1958 model was restored by Walter Bieri of Avenue City, Missouri.

In production from 1956 to 1958, the 230 was nearly identical to the 200 except for a higher compression ratio and white trim. You can also see the similarities to the original Farmall C.

A New Direction

Making its debut in 1958, the Farmall 560 was one of a new line of big tractors that featured new styling, new amenities, and a new level of power.

Just as Farmall tractors dramatically changed in appearance in the late 1930s with the introduction of the Raymond Loewy–styled tractors, so was the case in 1958. The revolution started with the Cub, 140, and 240. The tractors themselves weren't that much different than their predecessors, but they were vastly different in appearance, sporting new square grilles and performance improvements that included 12-volt electric systems, a new operator platform, and live hydraulics.

Other tractors to follow included the 460, 560, and 340, which featured a totally new design. The 560 was the first of Farmall's big tractors and, as such, featured a six-cylinder engine that was powered by gasoline, LPG, or diesel. Both the 460 and 560 also included power steering as standard equipment, signaling the start of even more emphasis on operator comfort and control. Unfortunately, the 460 and 560 were a public relations nightmare. Both tractors used variations of a 263-cubic-inch six-cylinder engine for more power; but the final drive remained basically unchanged since the 1938 Farmall M, which only generated half the horsepower. Consequently, it didn't take long before the powertrains began to fail. In the end, International Harvester spent thousands of dollars redesigning the final drive and replacing damaged ones. Worst of all, it occurred at a time when John Deere was passing IH to gain the number one position in agricultural equipment sales, which eventually happened in 1958.

While most collectors and restorers limit their pursuits and efforts to tractors built prior to the 1960s or even prior to the beginning of the square-grille models, International Harvester did continue to use the Farmall name for several more years, finally dropping the nomenclature in 1973. In the meantime, models that took the Farmall name into the mid 1960s included the 404, 504, 706, and the new powerhouse 806, which was billed as the world's most powerful all-purpose tractor, thanks to 94.93 hp from its six-cylinder 361-cubic-inch diesel engine.

One of the last IH tractors to carry the Farmall nameplate on the side was the 1206, which became the first IH model to surpass 100 horsepower. The 1206 was also the first to use a turbocharger as standard equipment.

Although later tractors were often referred to as Farmall models, the new 56 Series that followed, beginning in 1967, actually carried the name International on the side. Nevertheless, the lineage continued through 1985 with the 544, 656, 756, 856, 1256, and 1456, followed by the 826, 1026, 666, 766, 966, 1066, 1466, 1566, 1468, 1568, Hydro 100, and Hydro 70.

In 1984, after a series of financial setbacks, the Harvester Agricultural Equipment Group was acquired by Tenneco and merged with J. I. Case to form the Case IH brand, ending an era that had essentially been launched with one tractor in 1924. The old Farmall plant was closed and the last International Harvester tractor rolled off the assembly line in May 1985.

The Other McCormick-Deering and International Tractors

Although the Farmall name was perhaps the best known of the International Harvester offerings, it was by no means the only tractor line offered by the company. The Farmall name, in fact, was only used as a badge on row-crop tractors. Throughout the glory years of International Harvester, the company also marketed a full line of standard-tread tractors, wheatland-style plow tractors, orchard tractors, and industrial models—sometimes at the expense of Farmall sales. As an example, IH only produced 200 Farmall machines during its first year of production, fearing that the new Farmall would hurt sales of its standard-tread 10-20.

Fortunately, the company soon discovered there was room in the market for both types of tractors. As a result, both the 10-20 and the 15-30 were updated in 1932 to become the W-12 and W-30. The latter used the engine from the F-30, while the W-12 was a standard-tread version of the F-12. However, IH discovered that even the W-30 wasn't big enough for Midwestern wheat producers, leading the company to introduce the W-40. Unlike the Farmalls, however, all three tractors carried the McCormick-Deering name, even though some earlier tractors bore the International name.

In 1940, following the introduction of the Loewy-designed Letter Series, IH extended the new look to the standard-tread models. In effect, the W-4 became a standard-tread version of the Farmall H and the W-6 was the equivalent of the M. The WD-6, of course, was the diesel version of the 36 PTO/belt horsepower tractor. In 1940, IH also introduced the W-9 and WD-9, which boosted drawbar horsepower to 47.06 and 48.45 in gasoline and diesel versions respectively.

After going through the Super Series stage—just as the Farmall tractors did, to become the Super W-4, Super W-6, and Super W-9—the wide-tread models followed in the steps of Farmall with the Number Series. In the meantime, though, International Harvester introduced a number of variations for specific roles, including models like the McCormick-Deering I-4, I-6, and I-6 Heavy Duty (Industrial); O-6, OS-6, and ODS-6 (Orchard); and the Super WDR-9 (Rice).

When numbers replaced letters, customers saw another unique change in 1954. The McCormick-Deering name was no longer used, having been replaced with the International nameplate. Hence, farmers were introduced to the International W-400, W-400 Diesel, W-400 LPG, W-450, W-450 LPG, W-450 Diesel, W-600, W-600 Diesel, and W-650 in diesel, gasoline, and LPG versions.

The heavy tillage tractors weren't the only IH models to be marketed without the Farmall name, though. In 1955, IH also introduced the International 300 Utility. Designed to compete against Ford's N Series, it was mechanically identical to the Farmall 300, but had a lower profile and straddle-type seating. Other tractors that followed in the utility line included the International 330, 350, and 350 Diesel.

Even as tractors got larger and merged more heavy tillage and row-crop capabilities into one unit, IH continued to serve the utility market, unveiling a new line in 1971. Built in the Doncaster, England, factory, they included the 32-hp 354, 40-hp 454, and 52-hp 574.

Today, many of the early non-Farmall models continue to be highly sought after collectibles, particularly the rare units like the industrial, orchard, and rice models. Part of the reason, of course, is the low production volume. But just like all collectors, tractor enthusiasts are always looking for something different from what someone else has. And some of the non-Farmall models were truly different.

The W-40 was one of International Harvester's largest standard-tread tractors for its time. Note that standard-tread models still carried the McCormick-Deering name, while the Farmall badge was reserved for row-crop models.

Rated at approximately 50 belt horsepower, the W-9 was a popular tractor among wheatland farmers.

Orchard models, like this O-6, are the envy of any I-H collector, especially when the sheet metal is in this good of condition.

Built from 1940 to 1952, the W-6 was designed for smaller grain farmers who didn't need a row-crop tractor.

Weighing nearly 6,000 pounds without fuel or water, the W-450 was available in gas, LPG, and diesel versions.

Right: International Harvester's Model 300 Utility was the company's first official entry in the utility market, designed to compete with Ford's popular N Series.

Below: Like its row-crop counterpart, the 450 Utility was available in gasoline, LPG, and diesel versions. This 1958 model is owned by Rollie Moore, from Oneida, Illinois.

Above: The 460 Utility model was first produced in 1958 and included row-crop, wheatland, and utility versions, as well as an orchard and grove model. The diesel model featured a new 236-cubic-inch, six-cylinder engine.

Left: Case IH recently revived the Farmall name for a new line of compact tractors. Ironically, these new compacts are doubly as powerful as early Farmalls of the same size.

Chapter 2

Shopping for a Tractor

For some tractor restorers, shopping for a tractor is the adventurous part of the project. Where you start your search, though, depends a lot on your goals.

If you're buying a tractor strictly as a collector model, a whole different set of rules tends to apply. The rules depend, too, on whether you're collecting the model for its value on the market or its sentimental value to you alone. As an example, some tractor restorers simply have a desire to restore and collect all the models within a certain series, such as the Letter Series models or all the F Series. Perhaps you just want to own a model like the one you drove while growing up on a farm or visiting a grandparent. If that is the case, it's better to evaluate the tractor as if you were purchasing a model for doing work around the yard or farm. Since many of these models are not rare, it's best to look for a combination of sound mechanics, good cosmetics, and a reasonable price.

On the other hand, if you intend to buy a vintage tractor as an investment, you should plan on doing some research and perhaps checking out the credibility of the seller and model being represented. While most enthusiasts are honest people who share the love of Farmall tractors as much as you do, there are people who will intentionally or unintentionally misrepresent the products they have for sale. If you're interested in a particular model, study factory literature or tractor books to find out how many of that particular model were built. This will give you an idea how rare that model is and what it might be worth when you are finished. Find out, too, if there are any distinguishing characteristics of the tractor that might identify it as being the real thing—even if sheet metal or certain components have been changed.

Chris Pratt, with *Yesterday's Tractors* on-line magazine, explains that working with rarities almost always rules out looking for perfect mechanical and cosmetic condition. "I have seen extremely rare tractors purchased that consisted of just the engine block, rear end, rims, and frame assembly," he says.

On the other hand, Pratt cautions against buying a tractor on which the cosmetic components are the only thing that makes that specific machine rare. "A common example of this is some orchard model tractors," he explains. "Frequently, there are no remnants of the orchard add-ons or anything but a model designation to distinguish the machine from its common utility version brother. Finding orchard models may be relatively easy, while finding the orchard components that make your project collectible is next to impossible. If an incomplete model is priced as a rarity, it may be wise to pass."

T. W. Cook, a Farmall collector from Georgetown, Texas, who specializes in H models, says it's also important to check the sheet metal on a regular model. "Make sure it's all there," he says. "The little access panel on the grille, where the cultivator steering mounts, is almost always gone. And those are really expensive, running $50 or more to replace."

If you're shopping for a collector tractor, you should also take a look at the tractor's serial number. As a general rule, the lower the number, the greater the tractor's collector value. Interestingly, International Harvester began the serial number series on every new model at 500. Consequently, you can be assured that any F-12, F-30, A, M, or 300 with a serial number of 501 is more valuable than the average model.

A high serial number, on the other hand, might indicate that the tractor was one of the last models of its type to come off the assembly line. Of course, it also helps to know some history of the serial numbers assigned to the model you're inspecting. It would be good to know, for example, that production of the F-14 didn't start with serial number 500, but rather 124000. Because the F-14 replaced the F-12, IH simply took the last F-12 serial number, rounded it to the nearest thousand and continued on.

Normally, the first ten and the last ten units within any particular model run have more value than the rest of the units within that model run. That would mean that any F-12 in the range of 123940 and any F-14 below 124010 should be worth above average in price.

If a tractor is not in running condition, as is the case with this ancient F-20, you almost have to start your bidding price based on its junk value.

How to Tell Them Apart

To the inexperienced tractor enthusiast, some of the early F Series tractors look alike or very similar, especially when the only decals that identify a model F-12 or F-14 have long since worn away. Unless you can look at the serial number, it can be difficult to tell them apart, especially on models that were essentially the same size. Since the F-20 followed right behind the Regular as an improved version of the latter, it can be rather hard to tell those two models apart, as well.

For starters, unless they've been modified, the F-20 has an upward-mounted exhaust, whereas the Regular had the exhaust under the manifold. Still, you might want to look for more clues, because a number of Regulars were refitted with F-20 vertical exhausts over the years. One would be the steering gears at the top of the bolster, which were open on most Regulars. By the time the F-20 was introduced, IH had developed an enclosed housing.

When it comes to telling the difference between the F-12 and F-14, however, the best indicator is the position of the breather. You'll find it on the right side of the engine on an F-14 and on the left side on an F-12. The F-14 also had a raised steering wheel connected to a shaft that extended to the bolster at an angle. However, this isn't always a reliable indicator, since some F-12 owners retrofitted the change to their tractors, as well.

Finally, many novice restorers find it difficult to tell a Farmall Cub from the Farmall A, since both shared the Culti-Vision configuration and the Cub is only 20 percent smaller. You have to have the two tractors setting beside each other to tell that, however. One clue to the difference is the grille: on the A, the grille shares its appearance with the other Letter Series models. Another clue is the fuel tank. The Cub has a rounded tank, whereas the A featured a teardrop shape.

Some antique tractor shows have begun featuring a vintage tractor auction, which is one source for a restoration project.

Tap into the Resources

You certainly don't need to go it alone when shopping for a vintage IH tractor, especially if you're new to the collecting world. There are a number of resources available for help, including bulletin board postings on tractor websites, chat rooms, and collectors who exhibit their machines at antique tractor shows.

Another valuable source is the Wisconsin Historical Society, which maintains a vast McCormick-IHC collection. Comprised of approximately 12 million pages and items donated from International Harvester, Navistar, and Case IH, the collection offers a wealth of knowledge for Farmall and McCormick-Deering collectors.

According to Lee Grady, the McCormick-IHC collection archivist, the museum has more than 50,000 IH owner's manuals covering every tractor and implement built by International Harvester between 1902 and 1952. The collection of manuals after 1952 is still impressive, but it doesn't include every machine.

Grady says the collection also includes serial number records that list both the month and year of manufacture. Unfortunately, the archive was never given records of where individual tractors were shipped or sold. Still, when presented with a tractor serial number, Grady can generally tell you in what month it was built. That information, in turn, can help you determine what paint, accessories, etc., were appropriate for the model you own or are considering buying.

To further assist you in your research efforts, the collection includes thousands of original photos of vintage Farmall, McCormick-Deering, and International tractors, as well as color ads, posters, etc. Many of these are viewable on the website, and can serve as a reference for paint color, attachments, or equipment and decal placement.

Finally, the McCormick-IHC Collection has the records from numerous committees within the International Harvester organization. This includes all the paint committee decisions from 1924 through 1957; minutes from naming committee meetings in which decisions were made about product names; and records of the new works committee from 1910 to the 1940s. Other valuable information in the collection includes a 1940 farm equipment paint chart that shows 16 original

The spelling on this sign may not be correct, but at least potential buyers know where this tractor stands mechanically.

For sale at a tractor show, this W-4 lacks little for a complete restoration.

IH colors; 66 reels of microfilm of engineering drawings for tractors built before 1939; and answers to a list of frequently asked questions.

For information requests, it's recommended you send your inquiries in writing via fax, mail, or e-mail. The contact information is listed in the appendix. Grady says the only charge for services is for the cost of any photocopies requested. Before contacting the archive, however, you might want to do a thorough search of the archive's website. Grady says more information is being added all the time to give Farmall enthusiasts easy access to the collection.

Buying the Right Tractor

The time you spend looking for the right model and inspecting each tractor will pay off later in the form of greater efficiency, less time and money spent on restoration, and increased satisfaction with the finished product. If it's a work tractor you're looking for, the benefits may also include increased safety; buying a tractor that is not ideal for your needs may be not only inefficient but also dangerous.

Another aspect to consider when looking for a prospective project is the geographic location where the tractor was used. You'll quickly find that, as a general rule, tractors used in the eastern and southern parts of the United States are more apt to have rust problems or a stuck engine, while those from the western part of the country tend to suffer more tire damage due to sun exposure and dry rot. Humidity has a tendency to take its toll, and nothing is worse than salt air.

Larry Manville, a Farmall enthusiast from Faucett, Missouri, agrees that the dryer climate prevalent in the continent's High Plains helps preserve the classics. Nearly half of the vintage models he and his sons, Mark and Gary, have restored to date have come from farms in California. He insists that, in addition to having less rust, western tractors exhibit less wear on the steering mechanism and front axle, simply because the fields are bigger and flatter and the tractor has made fewer turns.

It's helpful, as well, to know the history of the tractor you're buying. For example, if the tractor was

Above: Any high-crop model is highly collectible. In fact, some collectors spend years scouring the country to find a model like this high-crop MV.

Right: Missouri tractor restorer Larry Manville traveled all the way to California to find these two rare high-crop models.

used to haul or load manure in a livestock operation, you can expect to spend some money on front wheel bearings and seals. Likewise, the steering mechanism on a row-crop tractor is likely to require more work than that of a wheatland version.

There are other things that can tip you off to potential problems. One is the hose between the carburetor and air cleaner tube: if it is cracked or missing, the engine may have sucked in a lot of dirt and will need work. Likewise, a missing or cracked shift lever boot can let water into the transmission—and nobody has to tell you what kind of problems water can cause when it freezes and thaws repeatedly.

In contrast, antifreeze in the cooling system and oil in the oil-bath air cleaner tells you right away that the tractor was treated well and is likely in good shape.

If the tractor is in running condition, there are several things you should check both before and after you start the engine. In fact, you may want to turn to chapter 5 on Troubleshooting and run through some of those procedures before you make your final decision.

If the tractor is not in running condition, you'll want to make sure the pistons are not rusted to the cylinder walls. One way to do that is to pull one or more of the spark plugs and shine a light inside the cylinder walls to check their condition. If the walls are shiny, the pistons are probably not stuck too badly, if at all. On the other hand, if you can't even get the spark plugs out due to rust, you might have cause for concern.

With its offset Culti-Vision platform, a Farmall B can be collectible on its own. But it becomes even more so with a single-wheel front end.

Negotiating a Price

Before you start dealing on any tractor, you should know your needs, your budget, and what is on the market. Become as knowledgeable about the prospective tractor as you can through research, conversation with other collectors, and physically checking it out. Here again, Chris Pratt and his *Yesterday's Tractors* website can be of value. The website includes a listing of recent tractor sales and lists a variety of models, their condition, and how much they sold for.

This research will allow you to go into the negotiations with a price in mind. If your preview of the tractor turned up any problems, you may find that the seller is willing to come down on the price. However, you need to decide if you have the time and expertise to correct what you have found.

You may find that the tractor won't start or run on the day you look at it. The seller may, in all honesty, tell you that everything worked fine when he or she last drove it; but when a tractor sits for long, it can develop problems of which even a careful owner is unaware. In this case, you must start your bidding from nearly scrap level prices, since you have no idea what you're getting into, then go only as far as your conscience and experience will allow.

Many collectors also go into a negotiation with the idea that a stuck engine is basically junk. If it can be freed, that's a bonus. Hence, your offering price should reflect that possibility.

The serial number plate alone can add value to a tractor—especially if it proves that the machine was one of the first or last produced in the series.

International Harvester had a unique serial numbering system that included a prefix, serial number, and suffix. With the proper codes, the number will tell you a lot, including the model, what year the tractor was built, what kind of fuel it burns, and whether it had rubber tires or steel wheels when it left the factory.

The engine on every IH tractor has its own serial number, which may or may not match the chassis serial number. In addition, the number may be on a plate or stamped on the block, as shown here.

Finally, know how much you are willing to spend on the whole project before you start negotiations. Many people don't realize how much expense is involved in a restoration. Even if you get the tractor at a decent price, you have to anticipate the cost of engine repairs, body work, new wiring, replacement components, and so on. A few things to consider that will add the most cost to your restoration project are tires (figure close to $800 if you have to replace them), and the condition of sheet metal, which can be costly if you have to purchase reproduction parts and fenders.

Many restorers suggest that you assume the worst of any potential restoration. That way you won't be surprised if you can't find parts or if it takes more time and money than you expected to get the job finished.

Experienced Farmall restorers say it's important to inspect closely the engine block on F-12 and F-14 models. The thin-walled block is known for developing hairline cracks like this.

Unique features, like the wide stance of this Model 200, add to the tractor's value among collectors.

T. W. Cook, a Farmall collector from Georgetown, Texas, warns that you need to make sure the cover plate that provides access to the cultivator steering mounts isn't missing, as is the case with this tractor. They can be costly to replace, he notes.

One way to distinguish the F-14, top, from the F-12, bottom, is to look at the steering shaft. The F-14 featured a raised steering wheel for greater operator comfort.

When shopping for a tractor, it helps to know what was original equipment. Obviously, this high-back, padded seat wasn't an original feature on the Farmall C when it was introduced in 1948.

Regardless of the model, LPG tractors were less common than gasoline versions, which makes them more collectible in many cases.

With more tractors than he could get to, Francis Gordon of Agra, Kansas, put this lineup of F Series models up for sale at a local consignment auction.

A Farmall F-12 awaits restoration outside a shop owned by Francis Gordon.

Some restoration projects are no more than a pile of parts when purchased. Obviously, the price you're willing to pay for such projects should depend on the rarity of the model.

While some restorers specialize in Farmall row-crop models, others prefer the powerful old wheatland models, like this LPG Model W-9.

More and more restorers are looking for implements or attachments to add to their restored tractors, which makes this combination particularly appealing.

Special features— like these dual rear wheels found on an old F Series model—can add a tremendous amount of value to a tractor. In this case, it nearly doubled the price the tractor brought at auction.

Due to their small size, Farmall Cubs have always been popular with IH enthusiasts. This one, parked next to an Industrial Model A at a show, displays the similarity in size and features between the Cub and its larger siblings.

It's always best to stick with something you like. Located in the heart of the wheat belt, Francis Gordon has a fondness for standard-tread and wheatland models.

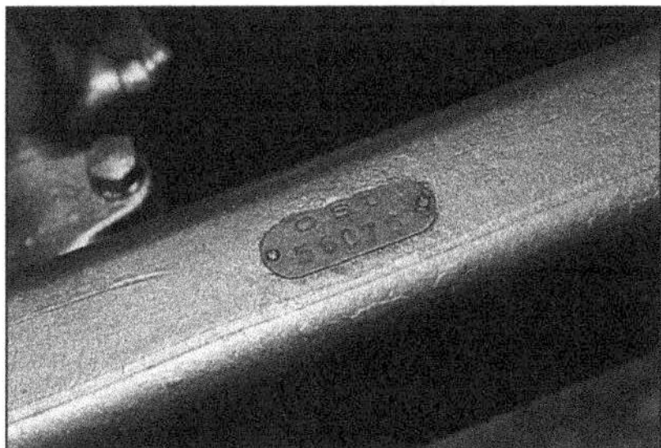

A Farmall Regular, owned by Paul Cummings of Amsterdam, Missouri, was not only one of the first 24 Farmall tractors ever built, it was once owned by Ohio State University, as evidenced by this brass tag on the frame.

Chapter 3

Setting Up Shop

Basic Tools

If you're serious about doing a first-class tractor restoration, the first thing you're going to need is a good set of tools. And make sure you put the emphasis on *good*. Cheap tools are only going to lead to frustration as they break, strip, or, worse, damage a tractor part. Look for a set of automotive-quality tools that come with a warranty, such as those offered by Sears Craftsman, NAPA, and Snap-On.

You'll want to start with a drive socket set that contains sockets ranging from 1/16 inch to over 1 inch. In addition to a ratchet handle, you'll need a breaker bar to loosen stubborn bolts without risking damage to the ratchet.

A set of combination wrenches will come in handy, too. There are some places you simply can't get a socket and ratchet into. You can decide which will work best for you and your budget, but choices include open-end, box-end, and wrenches that provide an open-end configuration on one end and a box-end of the same size on the other.

To round out your tool collection, you'll want to add a couple of adjustable wrenches (often referred to as Crescent wrenches, even though Crescent is a brand name), a full set of regular and Phillips screwdrivers, a pair of adjustable pliers, needle-nose pliers, and a pair of locking pliers (often referred to by the popular brand name Vise-Grip). Other tools that you'll probably need at some time or another include a good hacksaw, a punch set, and a cold chisel. And don't forget to pick up a couple of putty knives. You'll need those for scraping away grease and grime.

Don't assume you have to go out and buy all new tools. Due to the economy and changes in agriculture, farm auctions are far too common these days. So keep your eyes open for an estate sale or a shop liquidation where you can pick up what you need at a reduced price.

While your work area doesn't have to be anything fancy, it does help to have a place where you can keep the tractor indoors while it is being restored, especially when it comes time to paint the machine.

As a longtime farmer, Paul Cummings of Amsterdam, Missouri, has a well-equipped shop that serves as a site for restoration projects during much of the winter months.

Paul Cummings not only has tools conveniently arranged, but they're in a cabinet that can be locked.

Specialized Tools

Depending upon how much engine or electrical work you get into, there are other tools that you may need. These include feeler gauges, a point dwell meter/tachometer, voltmeter, and compression and vacuum gauges. For tools and gauges that are only used on occasion, consider renting or borrowing from a fellow restorer.

Other tools that can make your life a lot easier include a gasket scraper, bearing and hub pullers, and a seal puller. A pickle fork can be a handy item, too, if you plan to separate the tie rods on the front axle.

If you anticipate doing much engine work, you'll also need tools for making precise measurements. At a minimum, these should include a micrometer for measuring items up to approximately an inch in width; a dial caliper to determine the acceptability of parts such as the crankshaft and camshaft; a set of feeler gauges; and a dial gauge for measuring certain types of end play.

You'll also need a torque wrench. While most parts of the tractor don't require a specified torque rating, you'll find that many engine components, including the head bolts, must be tightened to a particular setting, in order to reduce the chance of head warping and oil and water leaks. While there are several different types of torque wrenches, perhaps the easiest to use is the type that makes an audible click when the correct torque rating has been reached. Since you preset the desired torque with a dial on the wrench, you don't have to worry about providing enough light or room to read a scale or being able to see the scale.

Of course, you can get by with the older style of torque wrench that uses a stationary pointer and a gauge attached to the handle. As the bolt or nut is tightened, the handle bends in response to the applied torque. As a result, the needle, which is fixed to the socket head, moves up or down on the scale to indicate the amount of torque being applied. You just have to be able to watch the scale as you're tightening the fastener.

Depending upon how far you get into engine repair, you may also need specialized tools such as a ridge reamer, valve-lapping tool, piston ring compressor, and cylinder hone. The use of each of these items is discussed in more detail in chapter 6, "Engine Repair and Rebuilding."

An oxy-acetylene torch can be equally valuable for cutting metal and heating stubborn parts that refuse to budge.

An angle grinder or stationary cutoff tool can be a valuable asset for sheet metal work.

A number of specialty tools—including items such as feeler gauges and valve lapping tools—can be found at a reasonable price at your local automotive parts store.

You may or may not need a welder during your restoration project. If you do need welding but lack the equipment or skill, you can probably find a friend to help you out or take the piece to a commercial welder.

While it's not feasible for the casual restorer, Paul Cummings has his own inventory of commonly used bearings and seals.

Paul Cummings also maintains an inventory of wheels that have already been cleaned and primed, so they're ready to replace one that's bad.

A torque wrench is just one of those essential specialty tools you'll need to purchase or borrow when overhauling a tractor.

Air Compressor

While an air-powered impact wrench can be a valuable asset when removing stubborn bolts, air-powered tools aren't quite as necessary as the air compressor itself. You'll want a portable air compressor with a tank for a couple of reasons. First of all, an air hose and nozzle are invaluable for blowing dust and dirt out of crevices and away from parts. You'll want to use it to blow out fuel lines, water passages, and the like, as well.

Secondly, assuming you're going to be painting the tractor yourself, you'll need an air compressor to operate the paint sprayer. Here, tank capacity is important. If the tank doesn't have enough capacity and the pump can't keep up, you're going to be painting a few minutes, stopping to let the pressure build in the tank, painting a few more minutes, and then waiting again. Hence, you might want to put things in reverse order and shop for a paint sprayer before you look for a compressor.

Most restorers who do their own painting suggest using a compressor with at least 1/2 horsepower that is capable of delivering at least 4 cubic feet of air per minute at 30 psi pressure.

If you plan to do your own painting, an air compressor with adequate capacity, an in-line water filter, and plenty of hose will make the job a lot easier. You'll find compressed air equally helpful when cleaning parts and components.

Vise

You don't have to own a vise to do a tractor restoration, but considering the availability and reasonable cost, and the versatility that a vise provides, you'll likely find it worthwhile. Just being able to clamp a part in the vise while you work on it can be helpful at times. And you'll be especially glad you have one when you've got a part in which a bolt absolutely won't budge.

For the most versatility, many shop owners recommend at least a 6-inch vise that is bolted securely to a solid bench. You may even want to get a piece of plate steel to attach to the bottom side of the bench for extra strength and support.

Anvil

Another tool that you'll at times find invaluable—especially if you have to straighten sheet metal—is an anvil. You don't have to invest in a commercial shop anvil, though. For what you'll need most of the time, a 2- or 2½-foot piece of railroad track rail will do. In fact, the rounded edge of the rail will work better than a real anvil for some metal fabrication.

If you occasionally need a flat surface or a square end for bending, you can weld a piece of bar stock across one or both ends of the rail so it will stand upright when turned rail side down.

Hoists and Jacks

Last, but certainly not least, you're going to need equipment to lift and support the tractor, engine, and other components. Naturally, your needs will depend to some extent on the type of tractor you're restoring. Some models, including the Cub, A, B, C, Super A, Super C, 100, 200, 130, 140, and 240 use the engine as a load-bearing member; in other words, there is no full-length frame. The front of the tractor attaches to the front of the engine, and the rear of the engine attaches to the transmission. To remove the engine for an overhaul, you have to split the tractor in half. That means you'll need a floor jack or bottle jack to support each half of the tractor.

The majority of Farmall tractors, however, employ frame rails that attach to the transmission. The engine then sits within this frame, which means that the entire engine can be lifted out while the tractor remains on its wheels. The W-6, W-9, 600, etc., feature a similar arrangement, except the engine sets in a bathtub-type frame that is bolted to the transmission. The front axle is connected directly to the bottom of the frame. This type of arrangement, however, generally requires an overhead hoist or an engine hoist to lift the engine out of the frame.

There will come a time, too, when you need to remove the rear wheels, front axle, and front wheels for cleaning, restoration, and painting. Again, you'll need some heavy-duty lifting equipment capable of raising the tractor to the level where you can block it up on stands or wooden blocks. Don't try to get by with concrete blocks! They can crumble or crack too easily, posing a physical danger. Don't try to pile blocks up too high, either. One option is to build cribbing under the frame, which means you place strong wooden blocks log-cabin style under the tractor or axles as structural support until the wheels can be safely reinstalled. Thanks to the number of holes in the side rails, which could be used for mounting equipment, cultivators, etc., a lot of restorers simply build a set of stands and

Thanks to the number of mounting holes in the Farmall frame, it's fairly easy to make a set of stands to support the front end of the tractor.

If you're planning to split the tractor for transmission or clutch repair, a set of stands, like this one offered by Omar Associates, can be an invaluable asset. A set of wheels on the rear half allow you to roll the two halves apart.

bolt them directly to the frame rails.

Some restorers like to block up the entire tractor from the start, pull the wheels, strip the tractor down to the frame, and work on restoration from the ground up. Others like to leave the tractor on its wheels as long as possible, work on components as they go, and roll the tractor out of the way when necessary.

The bottom line is your lifting and cribbing needs will depend largely on the size and type of tractor you're restoring and how you prefer to work. Just be sure that you keep safety in mind and that you have the right equipment to do the job. Don't try to lift the rear end of the tractor, for example, with a single bottle jack, even if it is an 8-ton jack. That's not what you would call adequate stability.

Purchase a Good Shop Manual

Considering the number of different Farmall, McCormick-Deering, and International tractor models built between 1924 and 1974, and the number of sub-models within each model, this book cannot go into detail on each and every model—especially with powertrain restoration. There is simply no way to cover all the specifics and idiosyncrasies. You'll also need a source of specifications such as tolerance limits, torque settings, and wear limits. Therefore, you need to purchase a repair manual for your specific model.

There are a number of good sources for service and repair manuals listed in the appendix. Intertec Publishing, for example, offers a complete line of their I&T (Implement & Tractor) Shop Service manuals, which are available for virtually all Farmall tractors you're likely to encounter. Jensales, Inc. in Clarks Grove, Minnesota, also offers service and parts manuals for virtually every Farmall and IH tractor that was ever built.

You might check with your local Case IH dealer as well. Depending upon the age of the tractor, an original manual may still be available.

Of course, if you want to pay the price, you can still find original service manuals for older tractors for sale by vendors at a number of flea markets, tractor shows, and swap meets.

Before you start a restoration, one of the first things you'll need to obtain is a good service/repair manual for your tractor model. You'll need it for tolerance limits and specifications, if for nothing else. A parts manual for your model can be equally helpful during reassembly.

Chapter 4

Getting Started

Take Your Time

Once you have it home, the first step in restoring a vintage tractor is to remind yourself that the process is going to take some time. Many of the people who restore antique tractors are retired farmers or mechanics who do it for the enjoyment of seeing an old tractor brought back to life. Others are full-time farmers who spend much of their winter working on a tractor. However, once spring arrives, their pet project tends to sit until ground preparation and planting are complete.

The point is, unless you're retired or have three or four cold winter months to devote to the project, you need to realize that a quality restoration is going to take up to a year or more. Trying to finish the project too quickly is going to lead to either discouragement or dissatisfaction later on with the shortcuts you have taken.

It can take a lot of blood, sweat, and tears to get a tractor to the condition of this beautifully restored Farmall AV.

Establish Your Goals

If you talk to many tractor restorers, you'll soon find that *tractor restoration* can mean lots of different things. And, indeed, one of the things you'll need to do right up front is decide how far you want to take the restoration project. To some enthusiasts, a vintage tractor restoration is nothing short of restoring the tractor to mint condition. That means they go through the engine, transmission, rear end, and every other component that might need attention. They also insist on accuracy in every detail and top it off with a quality coat of paint.

On the other hand, not everyone has the budget to do a first-class restoration. If you're in that category, you need to decide along the way what you can live with and what you can't. As an example, if you only plan to drive your finished product in a few parades a year and take it to a few antique tractor gatherings, you may not need to replace that gear in the transmission that is missing a couple of teeth. But if you plan on using the tractor to mow the roadsides, plow the garden, and push snow in the winter, you'll want to restore the transmission to like-new condition, or risk further, more costly damage.

Don't try to cut costs where it doesn't make sense, though. If there is one common lament among tractor restorers who make their living restoring tractors for paying customers, it's that some clients don't want to spend the money to do it right the first time.

Document the Process

As stated earlier, it may be a year or more before you get to the point where you're ready to reassemble the tractor. Consequently, it's important that you maintain good records as you disassemble the tractor. Keep a pad and pencil handy for recording measurements and taking notes. You should also consider taking pictures or shooting video as you go. Used in combination with a service manual, these images can be valuable several months down the road.

You'll want some good photos anyway, even if you don't need them for reference later on. One of the first questions people are going to ask you is "What did it look like when you started?" Showing them photos of the iron pile you dragged into the shop is half the enjoyment.

The first step in tractor restoration consists of stripping the tractor down and cleaning it up, starting with the layers of grease that have accumulated around the engine and transmission.

While you can still see where the decals were originally located, it's a good idea to take measurements before stripping and cleaning the panels. It'll make the job a lot easier and your work more accurate when it comes time to position new decals.

Finding Replacement Parts

Before you get too deep into the restoration project, you'll need to consider the challenge of locating and acquiring replacement parts. As you're disassembling the tractor for cleaning, begin making a list of all the parts you'll need to restore the machine to show or working condition. This will give you an idea of how much you may need to spend on parts like bearings, gaskets, sheet metal components, and so on. It will also give you a head start on locating some of those parts. By knowing what you need ahead of time, you can be searching the swap meets, salvage yards, and classified ads for the necessary components while you're working on other areas.

Record as much detail as possible, including part dimensions and their shapes and any serial numbers listed on separate parts. The good news is, parts for antique tractors are much easier to find today than they were just five or ten years ago, thanks in part to the growing interest in tractor restoration. Case IH dealers continue to stock parts for a number of tractor models that date back to the 1930s. In addition to original parts, there has been a proliferation of small companies that specialize in restoration parts. Through the sources listed in the back of this book and in many of the restoration and tractor club magazines, you can find everything from reproduction grill emblems and fenders to radiator caps and temperature gauges. There are numerous individuals and companies, too, that can repair your old magneto, carburetor, or distributor.

Keep in mind that the part you need might just be available from an unusual source. Some tractor dealerships, including most John Deere dealers, for example, can reference the number on any bearing, regardless of what tractor brand or model it came off of, and tell you in a matter of seconds, via the parts computer, whether they have a bearing available to match it.

Before you spend a lot of money on new parts, though, spend some time searching the salvage yards and used parts dealers for original parts that can be refurbished. The swap meets held in conjunction with a number of tractor shows are a good source of used parts, too. Not only will used parts make your tractor more original, but they may save you money. Just don't rely on used parts in critical areas where a failure could jeopardize safety or cost you more money later on. Breaking a seat spring , for example, can be downright hazardous, especially if you're pulling an implement. It's better to go with new or a reproduction.

While disassembling the tractor and cleaning parts, make a list of parts you'll need, so you'll have time to search sources and swap meets.

While some restorers use fine silica sand to strip the paint on sheet metal, others limit sandblasting to cast parts like wheels, engine blocks, and the frame. If you use a sandblaster near engine parts, though, make sure every hole that could allow sand to enter the engine is plugged, including water pump vents and so on. Also remove electrical components like the generator, starter, etc.

Clean It Up

Just the word *restoration* implies that the tractor you intend to bring back to life needs a lot of attention. More than likely, it is covered with fifty or more years worth of grease, dirt, and grime. And that's assuming it runs or has been protected from the elements for a good portion of its life. If you're restoring a tractor that has been sitting in the fencerow for the last twenty years, you're probably looking at a lot of rust, too.

Hence, the first step in any tractor restoration is trying to find the potential that is hidden beneath years of neglect. If you haven't picked up the tractor already, you can start by running it through a car wash on the way home, unless you have a hot pressure washer of your own. Naturally, hot water or steam cleaning is going to do the best job of removing oil and grime.

Bill Anderson, a full-time tractor restorer from Superior, Nebraska, likes to spray oven cleaner on some of the really greasy areas and let it soak before power washing the whole tractor. Others use a hand-pump sprayer to soak the tractor and engine with diesel fuel for several hours. Spray-on engine degreaser can be an asset, as well.

Chances are, though, it's going to take more than hot water and engine cleaner. Grease that has mixed with dirt and debris and become baked on by engine or transmission heat can get as hard as rock. So you'll want to keep a putty knife and wire brush handy while you're cleaning.

If you're doing a true restoration, you should start disassembling the tractor at this point anyway. Start by removing the major components, like the hood, fenders, fuel tank, and so on. Not only will it make the job of mechanical restoration easier, but you will be able to do a more thorough job of cleaning and painting the part if it is already off the tractor.

While cleaning the tractor and stripping paint, keep in mind that any traces of grease, oil, rust, and old paint can create problems with paint adhesion. The goal is to make sure every surface will hold the paint you will be applying later.

Naturally, there are several methods that you can use for removing grease, paint, and rust, and each has its own place, along with advantages and disadvantages.

It is not recommended that you use gasoline or kerosene as a cleaner. There are cleaners and degreasers available at any automotive store that are both safer and more effective.

Paint Removal

Once the tractor and all components have been thoroughly cleaned, it's time to start removing what paint is left. Before you do that, though, take note of any original decals that are still left on the tractor; this would be another good time to take pictures. Then grab a tape measure and a notebook and take notes on the position of each decal. You might note, for example, that the "L" on your Farmall decal should be positioned two inches forward from the seam where the hood and fuel tank meet; or the bottom of the decal is 1/2 inch from the bottom edge of the hood piece.

At this point, however, you may want to set the sheet metal parts aside, since they will be covered in

You'll be able to find everything from original parts that need work to reproduction parts and decals in the vendor section at many tractor shows.

more detail in chapter 15. In the meantime, you can proceed with paint removal on cast parts, the frame, etc. Of course, one of the fastest and easiest ways to remove old paint, rust, and dirt is by sandblasting. Just remember, there is a limit to where sandblasting can be used. The fine sand that works so effectively at removing paint has a way of working its way into cracks, crevices, and seals, as well. Sandblasting an engine or transmission, for example, can cause more problems than it solves, particularly if the sand is forced into the case or ruins the seals.

Consequently, it's important that, before you start any sandblasting, you go over the entire tractor, looking for holes that will let sand in and cause later damage. Many machines have passages to the brakes in the pinion housings. Be sure to pack these areas with heavy rags. Check, too, to see if any of the bolts you removed during disassembly left an open hole that will let sand into the clutch or shaft areas. If so, you'll want to put bolts back in these holes. Watch out for the clutch inspection plate, as well. It won't seal well enough to keep sand out. Most water pumps also have a vent on the bottom side that will let sand up into the bearing and shaft area. Don't count on just trying to avoid these areas, because if there is a hole, sand will get in.

Don't rely on masking tape to adequately protect an area, either. With enough pressure behind it, sand will go right through masking tape. Instead, some restorers rely on multiple layers of duct tape placed over any vulnerable components or openings. Similarly, if you don't remove the radiator prior to sandblasting, you might want to cover the radiator core with several layers of cardboard and seal the seams with several layers of duct tape. That is, unless you plan to replace the whole radiator core anyway.

Be sure you cover the serial number plate, as well. Depending upon the age of your tractor and the serial number, this can be a valuable component. Wheels and cast parts can be sandblasted without much risk of damage. In fact, if you're dealing with spoke wheels, sandblasting may be the best way to get in and around the individual spokes.

The last method of removing paint, and the one you're probably going to have to employ as well, is mechanical removal. Unfortunately, this method requires the most sweat and hard work. You'll find a wide range of "weapons" available at most hardware and automotive stores, but you might want to start with the basics, including wire brushes and putty knives. However, you'll find just as many low-cost "tools" lying around the house that are just as valuable. They include old toothbrushes, which are particularly handy for scrubbing delicate parts; cotton swabs, which can be dipped in paint thinner to clean hard-to-reach crevices; and pipe cleaners, which can be used to clean the channels and tubing. A wire brush or a sander that fits on an electric drill can also come in handy when removing paint and grease.

Restoring a tractor to original condition often means finding a source for antique tires. Miller Tire is one of the sources for odd sizes and rear tires with the original 45-degree tread pattern.

There are a number of materials on the market that can be used to repair metal parts that aren't subjected to a lot of pressure, including products like Liquid Steel and J-B Weld.

Disassembly

Either before, during, or after cleaning, you'll need to start tearing the tractor apart, beginning with things like the fenders, fuel tank, grille, hood, and so on. Again, take your time. Label parts, if necessary, so you can remember how they fit back together.

Obviously, you're going to have a number of bolts, nuts, and washers to keep track of, too. One way to organize them is to collect a bunch of egg cartons and put the nuts and bolts from different areas into individual egg compartments. You can even use and label separate egg cartons for different parts of the tractor (i.e., one for the grille and hood, one for the transmission cover, etc.). For bigger bolts or parts, you can use coffee cans or plastic butter tubs.

Whatever you do, don't throw anything away while you're tearing a tractor apart. Even if a piece is rusted beyond any possible use, it may be needed as a pattern for creating a new piece later on.

You'll also need to think ahead at times. As an example, some restorers have been known to leave the exhaust manifold attached to the head until they know for sure they can locate a replacement gasket.

Finally, be careful about using too much force when trying to remove rusted or frozen parts. In your haste to break things loose, it's easy to damage irreplaceable parts. Quite often, the best bet is to use a combination of penetrating oil, patience, and a proper tool. In other words, don't try to put a pipe or extension on a 3/8-inch socket handle and expect to break a bolt loose. It may take a 3/4-inch set or an impact wrench.

If the part can withstand the heat, a propane or oxy-acetylene torch and occasional taps with a hammer can be as effective as anything. Alternately using heat and penetrating oil can also be helpful. Just don't apply oil to hot metal or direct an open flame toward a pool of penetrating oil. Be careful, too, about using a torch on a part where the heat can be transferred to a bearing. Using a torch to loosen the flywheel on a John Deere two-cylinder tractor is a good example; unless you know you're going to be replacing the driveshaft bearing, the time you save may not be worth the cost.

In the early stages of tractor clean-up, a simple putty knife can be a valuable tool for removing baked-on grease.

Removing Broken or Damaged Bolts

There's a good chance that at some point during the restoration process, you're going to be faced with a bolt that has broken off during removal or had the head stripped to the point you can't get a wrench on it. On the other hand, you may have a bolt or fastener that simply can't be removed by ordinary means. So let's take a look at the alternatives.

One option is to drill a hole in the bolt, or what remains of it, and use an "easy-out" to back it out of the hole. Unfortunately, many restorers say easy-outs are more trouble than they are worth or cause more damage than if you just drilled the bolt out in the first place. "Easy-outs *aren't*," simply states one tractor restorer. "If you break one off, you have just created a bigger problem than you started with. The easy-out is made of some pretty hard steel, so once you've broken one off in place, you can no longer drill it out."

If part of the old bolt is still sticking above the surface, one alternative is to find a nut that is approximately the size of the broken bolt head. Place the nut over the broken bolt and then use an arc welder to fill the nut with weld, thereby fastening the nut to the bolt from the inside. In effect, you've created a new bolt head.

One of the last options, and the only one that is really effective for a bolt that can't be removed any other way, is to drill out the old bolt. Start by making a punch mark in the center of the bolt or bolt head. This will allow you to drill a starter hole through the center axis of the bolt. Now, begin drilling out the bolt using successively larger bits until the hole through the center is nearly as wide as the bolt. Be careful not to damage the threads in the parent material by using too large a bit or drilling through a hole that is off center. In some respects, this process isn't unlike that used by a dentist doing a root canal. Nice thought, huh?

One restorer says he has had good luck using a set of left-handed drill bits to drill out a bolt, once a pilot hole has been established. Since the bit is turning in the direction you want the bolt to move, it will often come out as the heat increases and the center is hollowed out by the bit.

Shaft Repair

One of the things you'll deal with most often in a tractor restoration is the replacement of seals, bushings, and bearings. Unfortunately, you may also run across the occasional shaft that has been damaged by a defective seal. This is generally evidenced by a groove in the shaft that is deep enough that you can feel it when you run your fingernail across it. In effect, replacing the seal at this point is not going to solve the problem. As long as the shaft is grooved, it's still going to leak.

The good news is you have a couple of options short of buying a new shaft. One is to sleeve the shaft with a sleeve such as a Speedi-Sleeve that fits over the original shaft to create a bridge over the groove. The Speedi-Sleeve is marketed by Chicago Rawhide, or CR for short.

To sleeve a shaft, you'll need to accurately measure the diameter of the shaft at a point where it hasn't been worn. Then, it's simply a matter of taking the measurement and the application information to your bearing or parts supplier. In most cases, the sleeve is thin enough that a different-sized bearing or seal is not required. The trick is just getting the sleeve installed on the shaft, using a special tool, since the fit is designed to be tight.

Never compromise on safety during a tractor restoration. An old seat spring, for example, should always be replaced if its integrity is in doubt. Having it break while the tractor is being driven could be disastrous.

Another technique for repairing a groove or nick in a shaft is to fill it with a metal-type epoxy, such as Liquid Steel or J-B Weld. Generally, it's better to apply two or three thinner coats and build it up, rather than one thick coat. After the compound has cured, the shaft can be sanded down until the repaired area is flush with the rest of the shaft. Just be sure to use fine-grain sandpaper or emery cloth to start, so you don't scratch the shaft. Finish off the sanding with even finer-grit paper, in the neighborhood of 600-grit, for a smooth surface on the seal or bushing.

Chapter 5

Troubleshooting

Troubleshooting should be just a supplement to your tractor-buying procedure. In other words, there are several steps that could be considered troubleshooting that you should have already done in the process of evaluating the tractor when you bought it. However, for the sake of finding out how serious the problems are, let's take a look at a few more things.

Whether you are an experienced tractor mechanic or a novice working on your first tractor, your senses can tell you a lot about what is wrong—if you know what you are looking for. Your sense of smell, for example, can tip you off to a problem with the radiator, clutch, or engine. Your sense of hearing can tell a distinct difference between a tick, a knock, and a grinding noise. And there can be dramatic differences in the causes of such noises.

Start by looking on and under the tractor for traces of fluid. If what you find is a lighter hue, has the consistency of light maple syrup, and lacks the burned smell of combustion, it is probably from a leaking hydraulic fitting. A darker shade of brown that collects under the oil pan should be obvious. Engine oil is somewhat thicker and has the characteristic smell of having been in an engine; often it will leak out of the engine seals or leaky oil pan gaskets.

Evaluating a Tractor That Runs

Some restorers insist that unless the engine is totally locked up, it's usually best to start up the tractor and see how it runs in order to best figure out what it needs. If that means overhauling the carburetor or the magneto first, or installing a temporary gas tank, that's what they will do. At least then, you can listen to the engine, run it through the gears to see how the transmission and final drive sound, and evaluate the various systems.

Still, Chris Pratt, with *Yesterday's Tractors* on-line magazine, insists there are a few things you can check to avoid surprises later on. The following inspection points are just part of his troubleshooting and evaluation procedure.

1. Cooling System Inspection

Whether you're checking out a tractor before or after you've made a purchase, it's understandable that you want to fire it up without further delay. However, it will be to your advantage to check out a few things while the engine is still cold. This includes the cooling system, oil pan, and transmission.

First, carefully remove the radiator cap and check for coolant. There should at least be some type of liquid in the radiator. The best thing you can find is an antifreeze solution. The next best thing is clear water. If you find only clear water, though, you have to wonder if it was added since last winter or if it has been in there awhile.

If you find rusty water, you can expect to find pitting inside the engine and possibly a radiator that

Oil leaks and the color of the grease or oil are obvious signs of a leak or seal failure.

is leaking or about to leak. Rust in the cooling system is often an indicator that the coolant has become acidic, which means it could also start attacking metal components.

Last, but certainly not least, you'll want to make sure the coolant isn't oily. That can be an indication of seal failures; cracked parts, which are allowing oil and coolant to mix; or pitted parts, which can do the same thing.

If you find that the system is leaking, look for bad hose connections in the cooling circuit, especially where the hose ends meet the radiator or connect to the thermo-siphon components. Also check the radiator core for cracked tubing or leaky ends. If leaks in this area are excessive, the core should be replaced, since the internal integrity of the core itself is probably not worth salvaging.

From the radiator, you'll want to move back to the water pump, if the engine is so equipped. Look for a steady but slow dripping or for antifreeze streaks down the front of the engine housing. Most water pumps have a hole at the base that will leak antifreeze and coolant if the seal on the pump is in need of replacement.

2. Oil, Transmission Fluid, and Hydraulic Fluid Inspection

Next, you should take a look at the oil. But don't just pull the dipstick to see if the level is where it should be. Take a wrench and carefully loosen the drain plug to the point you could pull it out if you weren't holding it in place. Now, back off just enough pressure to drain out about a cup or less of oil and check it for water and antifreeze. If you get pure oil, you can rest a little more comfortably, knowing that water will settle to the bottom of a cold oil pan or gear case. A small amount of water can simply be condensation and may or may not be cause for concern. But if you find antifreeze, it should raise a red flag concerning the mechanical condition of the engine.

Repeat this process for the transmission and hydraulic reservoir.

3. Engine Starting Evaluation

Okay, now you're ready to start the engine. Does the engine start easily when cold? Knowing that a tractor starts after just a few revolutions easily eliminates many of your concerns in one check. You may not have a guarantee on the condition of each component, but you immediately know that the battery, compression, ignition wiring, distributor/magneto, fuel flow, and carburetor are in reasonable condition.

If the engine doesn't start easily, you may still be looking at a good tractor, but you know there is a little more work ahead.

Unfortunately, if you're looking at the tractor for the first time before buying it, and the unit has already been removed from the shed and warmed up prior to your arrival, you've lost out on a piece of information—namely the cold start.

4. Engine Running Evaluation

Does the engine run well when it is hot? Taking the time to see how the engine runs after it has been warmed up is particularly important if you are going to be using the tractor to pull a load. Plan to spend at least a half hour running the engine to check for problems that can cause it to run poorly after it warms up. In the meantime, you can look for leaks in both the engine and radiator.

After completing your inspection, shut the engine off and see how easily it starts when it is warm.

5. Exhaust Evaluation

Check for smoke from the exhaust. Blue smoke often indicates internal problems, such as problems with rings, pistons, or valve guides that are difficult and costly to repair. White or black smoke can frequently be corrected with carburetion or ignition changes. However, white smoke can also mean that water is getting into the cylinders, possibly through a bad head gasket, or worse, a cracked block.

6. Engine Noise Evaluation

Listen for noises from the engine. A ticking from the top of the engine may indicate the need for a simple valve adjustment, while a clunking or thumping sound deeper in the engine could indicate serious or expensive repairs. If possible, check if the sound becomes more pronounced under load. If a clunk becomes louder when the engine is put under a load, it may be an indication of problems with the crankshaft, bearings, or piston rods.

7. Oil Leakage Inspection

After the engine has been running for a while, shut it off and check the oil for foaming or the presence of water. Either condition is cause for concern. Check for oil seepage from the head and look for structural cracks in the block. These procedures may be difficult to perform if the engine is encrusted in dirt and grease, but they can be time well spent. Check over the cast and steel components and look for hairline cracks.

8. Electrical System Inspection

There should be some charge showing on the ammeter when the engine is running. You should also see some change in the charging level when the lights are turned on. This indicates that the regulator or resistor switch and cutout are operating properly.

Checking the engine oil for traces of water, antifreeze, or foaming can tell you a lot about the condition of the engine and cooling system. Whitish oil often indicates water contamination in the engine.

There was no oil dipstick on older Farmall tractors like this F-12. You'll need to check the oil condition by opening one or both of the petcocks on the side of the oil pan. This characteristic extended well into the Letter Series.

Whether you start the engine or just turn it over, checking for spark at the spark plugs will tell you a lot about the ignition and electrical system.

9. Governor Inspection

At some point, you'll want to make sure the governor is working correctly. Its job is to adjust the throttle to maintain engine speed as the load changes.

One of the easiest ways to check governor, says T. W. Cook, a Farmall H enthusiast from Georgetown, Texas, is to go down and up a small hill in fourth or fifth gear (be careful in fifth as a tractor like an M or H will go up to 15 mph; and if the steering, brakes or tires are questionable, this might not be a safe procedure). As you start downhill from level ground, you should hear the engine quiet down as the throttle backs off. Then as you start back up the hill, it should automatically throttle back up to maintain the speed.

10. Transmission Check

Assuming the tractor is drivable, you should also check the transmission and final drive by driving the tractor around and listening to the gears. Cook says second gear in the letter series tractors was commonly used for plowing and is likely to be worn the most. If the transmission is worn, you'll probably hear a whine in second gear. As mentioned earlier in the book, this may or may not be a problem, depending upon how you plan to use the tractor.

11. Hydraulic System Inspection

If the tractor is so equipped, test the hydraulic system. Extend the rams on any hydraulic cylinder to test the range. If possible, use the system to lift a load, then let the load sit in the hold position for a few minutes to see if there is any leakdown. Chattering noises from the pump while lifting the load may indicate that the pump is getting an insufficient flow of hydraulic fluid. The pump will have experienced excessive wear if it has been run this way for long periods of time, and it may be on the verge of failure.

At some point—preferably before buying the tractor—you should check the water in the radiator for traces of rust or oil. The best thing you can hope to find is clean water with antifreeze. In this case, Gary Manville, of Dearborn, Missouri, found a skim of oil on the top.

Simple Fixes

On occasion, a simple fix is all that is needed to correct what may seem to be a complex or an expensive problem. Always check the simple things first to avoid spending time and money on restoration steps that may not be necessary. Through his experience with tractor restoration, Chris Pratt has come up with the following list of common problems and simple fixes for them.

Runs Poorly When Warmed Up

Before replacing the carburetor, check the fuel line, sediment bowl, and tank outlet. With old tractors, these often become clogged with rust sediment and cause the engine to run as if the float and float valve are damaged. Quite often the tractor will run fine when started, but begin to starve out and miss after awhile.

Dies When Warmed Up

If the tractor warms up, then suddenly dies with no spark, and you find that the spark does not come back until the tractor cools down, the problem is most commonly a bad condenser. Since testing condensers seems to be a lost art, it is simplest to replace them.

Good Battery Won't Actuate Starter

This problem may be most common on tractors with a 6-volt system. Before replacing the starter, check for warmth at the connections of the battery cables. It may be that the cables are of too high of gauge (the wire is too small) or the connections may be less than perfect. As a rule, 6-volt systems draw more amperage than a 12-volt system, and the connections and wiring need to be near perfect for the starter to function as it was intended.

The Engine Is Getting Gas and Spark, but Won't Start

If the engine is getting a spark at the right time and gas is getting to the plugs, yet the tractor won't start, it is likely that your gas has gone bad—particularly if the tractor has been sitting for some time. The solution may be as simple as draining and replacing the gas.

Won't Start, Water in Distributor Cap

If you have trouble with your tractor during high-moisture times, such as during a thaw or in damp conditions, check under the distributor cap for moisture. In most cases, all you have to do is dry it out and hit the starter. Since it displaces moisture on electrical connections, WD-40 sprayed on the inside of the distributor cap can also do the trick.

Overheating or Not Charging

Before you replace your water pump, thermostat, and radiator cap, be sure the drive belt is the correct width and profile. Also ensure that it is tensioned properly. These factors can also cause the charging system to appear to be faulty.

Boiling Out Radiator Fluid

If the tractor is boiling out radiator fluid every time it warms up, the first thing you should do before replacing the thermostat is make sure the radiator cap is rated correctly for the system and that its spring and seal are still in good shape. A faulty cap may be letting off steam under what was supposed to be normal pressure.

Burning Oil

What if a compression test indicated good compression on all cylinders, showing that the valves, pistons, and rings are in good shape, but traces of oil smoke are coming out of the exhaust? This can be caused by the oil-bath air filter. Be sure that you are running the correct weight of oil. If the oil is too light, it will be drawn into the engine. Don't go overboard the other way, however; if the oil is too heavy, it won't clean the air. To learn more about oil-bath air filters, refer to chapter 13 on the fuel system.

Compression Testing

If you've had a chance to start or drive the tractor, you probably have an idea how well the engine runs. But for a real test, before you make the purchase or start tearing it down, you should run a compression test. This measures the pressure built up in each cylinder and helps assess the general cylinder and valve condition. It can also warn you of developing problems inside the engine.

Before you begin, you should start the engine and let it warm up to normal operating temperature. Now, shut off the engine and open the choke and throttle all the way to provide unrestricted air passage into the intake manifold. Remove all of the spark plugs and connect a compression gauge to the No. 1 cylinder following the gauge manufacturer's instructions.

At this point, you need to either have someone else crank the engine or use a remote starter switch that has been connected to the starter relay. Always follow all manufacturer's safety instructions, and make sure the transmission is in neutral and the wheels are blocked or locked prior to engaging the starter.

Crank the engine at least five compression strokes or until there is no further increase in compression shown on the gauge. Remove the tester and record the reading before moving on to the next cylinder. In addition to the psi rating, you should also note whether the needle goes up all at once, in jerks, or a little at a time.

You'll need to check the service manual for your tractor for the recommended pressure, but generally the lowest pressure reading should be within 10 to 15 psi of the highest. Meanwhile, engine compression specs can vary anywhere from 80 to 150 psi. A greater difference indicates worn or broken rings, leaking or sticking valves, or a combination of problems.

"The specific numbers don't matter as much as them all being even," says T. W. Cook. "If you're seeing 80 or 90 across all four cylinders on a tractor like the H, that's a good sign. If one or more are significantly lower, you'll probably need rings or worse."

If the initial compression test suggests a problem, you might want to confirm your suspicions with a "wet" compression test. This is done in the same way as your previous test, except that a small amount of heavyweight engine oil is poured into the cylinder through the spark plug hole before the test. Since this will help seal the rings from the top, it should help pinpoint the problem.

For example, if there is little difference between the wet and dry tests, the trouble is probably due to leaking or sticking valves or a broken piston ring. However, if the wet compression reading is significantly greater than your first reading, you can assume the problem is worn or broken piston rings.

If, on the other hand, two adjacent cylinders have similar low readings during wet and dry tests, the problem is more likely a defective head gasket between the two cylinders or a warped head-to-block surface.

Your notes on how the needle moved up can tell you a lot, too. If the needle action came up only a small amount on the first stroke and a little more on succeeding strokes, ending up with a very low reading, burned, warped, or sticky valves are indicated. A low pressure buildup on the first stroke, with a gradual buildup on succeeding strokes, to a moderate reading can mean worn, stuck, or scored rings.

There can be good news, however. If the readings from all cylinders are within reasonably close proximity, you can assume that the upper end of the engine is in good condition and may not warrant an overhaul. A simple tune-up may suffice.

If you're checking the compression on a diesel engine, the process is basically the same, except you will need to remove the injectors and seal washers. Plus, since the compression is higher on a diesel engine, you'll either need to use a different compression gauge or an adapter for diesel engines.

A compression check measures the pressure built up in each cylinder and helps assess the general cylinder and valve condition. It can also warn you of developing problems inside the engine. When performing a compression check, you'll want to record both the pressure reading and the rate at which the pressure increased. You should also compare the pressure readings between cylinders. Squirting a little oil in the cylinder and performing a "wet" compression check will give you an idea whether a compression problem is caused by faulty valves or worn or broken piston rings.

Chapter 6

Engine Repair and Rebuilding

The first step in engine repair and restoration is to find out what you are dealing with and what repairs are necessary. If you're lucky, you are restoring a tractor on which the engine is already in running condition. If that is the case, all that may be needed is a good tune-up. Chapter 5 on troubleshooting should have given you some ideas about how much repair is necessary.

On the other side of the coin is the engine that is completely frozen. Or the tractor may have been parked and left to rust after a piston rod broke or the engine block cracked. In either of those situations, you better plan on a complete engine rebuild.

The more common situation, though, is the engine that will run or start, but performs poorly. Perhaps it smokes, or has a distinct knock. If you're like some restorers, you may choose to overhaul the engine as part of a restoration. And if you're like others, you may be on a budget that demands just fixing what needs to be fixed. Either way, this chapter will hopefully guide you through some of the processes.

Freeing a Stuck Engine

Ask a dozen tractor restorers how to free a stuck engine and you're likely to get a dozen different answers. Each one seems to have his own favorite method. Unfortunately, few of them are quick fixes; they all take time and patience.

Most often, engines get stuck because the pistons and cylinder walls or sleeves have rusted together. This can occur as a result of water directly entering the engine or by condensation, or "block sweat," inside the engine, which leads to flash rust.

With a little time and work, flash rust can often be broken loose fairly easily. Unfortunately, pistons that are practically welded to the cylinder walls are a much bigger challenge.

Before you attempt to break anything loose, though, there are a few things to keep in mind. First, be careful about towing the tractor in gear in an attempt to free the pistons. Even if you have soaked the pistons for some time, you run the risk of damaging the engine. One possibility is that you will bend the connecting rods or components in the valvetrain, should the pistons come loose but the valves do not.

You also need to be careful about putting too much pressure on any one piston—such as with a hydraulic press—if all the pistons are still connected to the crankshaft. If the piston on which you're pushing comes loose but three others are still stuck, you not only risk damaging the piston, but the connecting rod or the crankshaft as well.

One tractor restorer says he likes to soak the piston heads and cylinder walls while leaving the connecting rods attached and the tractor blocked up on one side and in gear. Then every few days, as he walks by the tractor, he gives the back wheel a push to see if anything has loosened up. If you do attempt to tow the tractor at this point, make sure you are using a slow speed and you're pulling it over soft ground or gravel so the wheels will skid. Letting the wheels get a firm grip is a sure way to bend something.

Soaking the cylinders and using a press or hydraulic ram to force them loose is another option. Another method practiced by some is to put a chain around the engine while it is still in the tractor and pour oil into the cylinders. A round, wooden block is then cut to fit the cylinder. Finally, using the chain as a brace, a hydraulic jack is placed against the block and pressure is applied to the piston until it comes loose.

From the looks of this cylinder block, you might expect at least one of the cylinders to be stuck in place.

When it comes to the type of lubricant or penetrating oil to pour into the cylinders, everyone seems to have their favorite recipe. Some simply use diesel fuel, while others prefer something as exotic as olive oil. Still others prefer a mixture of ingredients that may include brake fluid, penetrating oil, automatic transmission fluid, kerosene, Hoppe's gun solvent, oil of wintergreen, Marvel Mystery Oil, and Rislone. One restorer claims to have freed the pistons on fifteen different engines with a mixture of one-third automatic transmission fluid, one-third kerosene, and one-third Marvel Mystery Oil. Another prefers to use Seafoam, a solvent available in most automotive stores that is used for everything from cleaning carburetors to lowering the gelling temperature of diesel fuel.

Of course, if those solutions don't work, you can always try Coca-Cola, as some would suggest. Although they recommend the diet variety to keep sugar from gumming things up, they insist it is as effective at cutting rust as anything on the market.

The important thing to remember when trying to free a stuck engine is that it took time for the engine to set up and it will take time to free it. If all else fails, remember that all Farmall and McCormick-Deering engines, except the Cub, use sleeves in the block. That means you have the option of destroying the piston in an effort to remove it, knowing that you're planning to replace the pistons and sleeves anyway.

Ring Job or Complete Overhaul?

If you have seen traces of blue smoke coming from the exhaust and the engine has been using quite a bit of oil, chances are pretty good that you are in the market for a set of piston rings. Before you can know for sure, though, you will need to check the piston-to-sleeve tolerances and surfaces, plus make sure the valve guides are not sloppy. The valve guides can exhibit the same symptoms as worn pistons and sleeves.

In most situations, it is not advisable to replace only the rings in an engine, because by the time you do the tear-down and measurement of the components, you'll find something else that justifies the need for a complete rebuild. If you are planning to replace only the rings, however, you need to first verify all of the following:

- Bore of cylinder is not scored
- Piston is not scored, cracked or its top surface is not eaten away
- Rings are not stuck to the cylinder wall(s)
- Bottom flanges of sleeve are not cracked
- Sleeves are not leaking oil into cooling system
- Bore is within tolerance throughout piston travel (up and down and across right angles around the bore)
- Piston ring grooves are within tolerance and not damaged

For a complete overhaul, the engine may need to be removed from the frame—or the tractor will need to be split if the engine block serves as a load-bearing member.

Regular, F-20, F-30, W-30, and 10-20 models have two access holes in the engine block that allow you to remove the pistons and connecting rods without removing the cylinder head. They also provided access for cleaning the screen on the oil pump without removing the pan.

Engine Disassembly

If you have a service manual for your tractor, it is best to follow the disassembly process outlined by the original manufacturer. In the absence of that material, though, the following procedure and tips should apply to most tractor models.

First, if you haven't done so already, you will need to drain all the fluids from the engine. This includes water and antifreeze, fuel, and oil. You'll also need to remove the engine hood, grille, radiator, and fuel tank to expose the engine.

Now, you may want to move to the bottom of the engine and remove the oil pan. This is fairly straightforward in most cases, but if the oil pan is made of cast material, you should loosen the bolts in an alternating pattern to prevent any warpage. As you remove the oil pan, look for any pieces of metal or shavings that might be a clue to engine problems.

At this point, you have a decision to make. If you're only replacing the piston rings, it is possible on some models to pull the pistons out from below by simply removing the rod bearing caps. Unfortunately, if you don't remove the head, you will not be able to remove the ridge at the top of the cylinder. This process can only be done from the top with a tool called a ridge reamer.

This ridge is formed naturally as the piston and rings travel up and down thousands of times in the cylinder. In essence, the area of piston travel wears, while the portion above it does not. If you don't remove this ridge, though, there is a possibility that the new, larger, sharp-edged rings will be broken by contacting the ridge. This means another tear-down, another set of piston rings, and the chance of sleeve replacement due to scoring. So for the sake of doing it right from the beginning, it's usually best to go ahead and take off the head. This will also let you remove the pistons from the top; plus, you can test the valves for proper sealing and measure the tolerances of the head and block surface.

First, remove any components or accessories attached to the engine, such as the coil, spark plug wire brackets, etc. Also remove the valve cover. If the engine utilizes a valve-in-head design, you should remove the intake and exhaust manifolds at this time, as well.

Now, begin loosening the head bolts a quarter turn at a time in an alternating pattern. If you have a service manual for the tractor and it shows a tightening pattern, simply loosen them in reverse order. This will slowly relieve the stress on the head and lessen the potential for warping or cracking.

If it becomes necessary to pry the head off to get it loose, make sure you are prying against the gasket, and not the block. You'll be replacing the gasket later anyway.

Once you have the head off the block, you can remove the ridge at the top of the cylinder using the ridge reamer. Since this is one of those tools that is only used on occasions, perhaps you can borrow or rent one.

Now, you can remove the rod-bearing caps and pull the pistons out the top of the block. Then, examine both the pistons and the cylinder walls or sleeves for scoring that would suggest the need for more than just ring replacement. In the process, try to keep carbon from getting into the cylinders, particularly the water and oil passages.

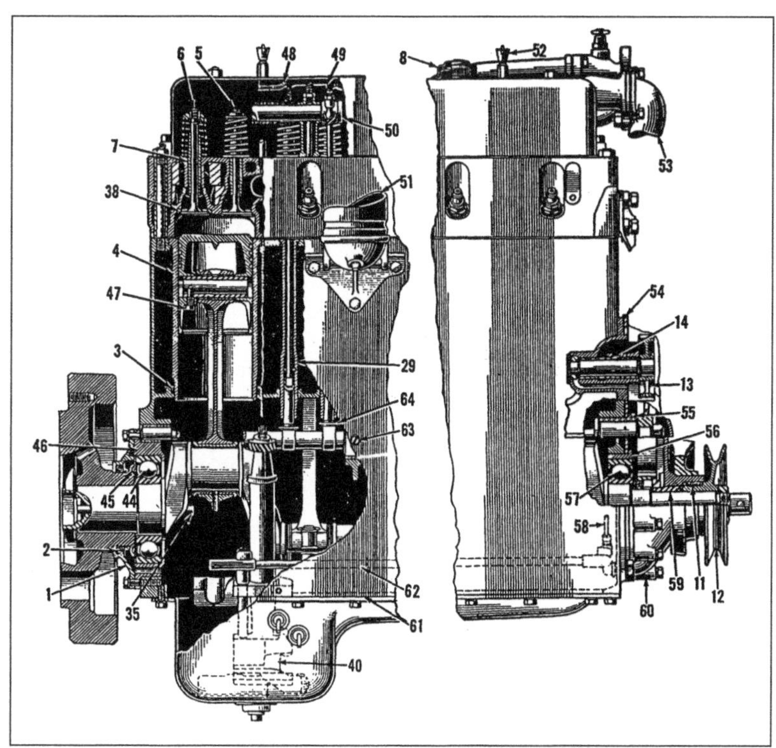

This side sectional drawing illustrates the cylinder block, pistons, valves, and related parts of the F-30 and W-30 engine.

Piston Ring Replacement

Once you have removed the pistons and checked them for scoring, you'll need to remove the rings by carefully spreading them from the break or ring gap. This is most easily done with a special tool called a ring spreader. An inexpensive ring spreader looks like a pair of pliers that open when squeezed. More-expensive ring spreaders have the same design but also have a band to wrap completely around the circumference of the ring to ensure that you don't elongate or spread the ring gap too far. This is not that important on your old rings since you are throwing them away, but on the new ones it is critical.

Once the rings are removed from the pistons, examine the grooves that the rings fit into and make sure they are not damaged. You will also need to carefully clean the grooves to remove carbon and dirt that would hamper the correct seating of the new rings.

Next, you'll need to determine the size of the new rings by measuring the bore and determining what oversize will completely fill the gap when the piston is at the top of its stroke. The manufacturer's required ring gap should also be taken into account. The ring gap is the clearance left at the split in the ring when the ring is as compressed as it will be in the cylinder. This usually occurs at the top of the piston's stroke.

In most cases, going one oversize up from the existing rings is sufficient, since ring replacement is done only when there is little wear on the piston and cylinder. If it takes more than one size increase, you might want to consider a more thorough overhaul.

Once you've determined the ring size, you'll need to hone the cylinder to remove the smoothness—generally referred to as a *glaze*—from the cylinder bore. Otherwise, the new rings will not seat properly. A cylinder hone that fits into a 1/4-inch drill can be found at most any auto parts store.

The intent of cylinder honing is to get a nice cross-hatch surface on the cylinder. This requires moving the hone up and down as the drill operates. Never allow the drill to run in one spot, and keep the hone lubricated and cooled with a fifty-fifty mixture of diesel fuel and kerosene, penetrating oil, or other thin lubricant. Be sure there are no large particles on the bore or hone surfaces that will cause scoring. Also, cover the crankshaft rod journals while honing to keep them protected from falling debris.

Now, it is time to check the ring gap. As stated earlier, the ring has to have the minimum specified compressed gap when it is in the cylinder bore to allow for expansion that occurs when the engine reaches operating temperature. Otherwise, the ring ends might butt together and cause scoring and ring breakage. Check your repair manual for the exact specification, but generally, it is considered to be 0.002–0.003 per inch of cylinder bore diameter.

To measure the gap, you'll need to compress the first ring and place it inside the bore. Don't put it on the piston. Now, push the ring into the cylinder using an inverted piston. This not only makes it easier to push the ring into the cylinder or sleeve, but it ensures that the ring is square with the cylinder wall. Take your feeler gauge and measure clearance between the ends of the ring. Compare this with the specifications in your manual and determine what changes, if any, are

While honing the cylinder, keep the drill moving to create a nice cross-hatch surface. Also be sure to keep the hone well lubricated.

necessary. Insufficient clearance will require that the ring gap ends be filed down to tolerances.

One of the best ways to do this is to take a file and mount it vertically in a vise. Take the ring and, holding it firmly in both hands, draw it downward over the stationary file. After removing a small amount of metal, check the ring in the bore again, repeating this process as often as needed. Follow this procedure for each ring making sure you note each gap specification for each particular ring.

After the gap for each ring has been established, take each ring, insert the edge of it into the corresponding ring groove in the piston, and measure the side clearance to determine if the ring grooves are worn. There should be ample room for the proper feeler gauge between the ring and piston lands. This will be another case where you need to check the manual for the specifications. Too tight a fit will keep the rings from proper rotation and movement as the piston moves up and down. Excessive side clearance, on the other hand, will allow the rings to flutter in the piston grooves when the engine is running. This can result in poor sealing of the combustion chamber or, worse, eventual ring breakage. Make sure you place each ring in corresponding order with each piston groove and check the ring for a mark indicating the correct side up. If the ring side clearance exceeds specifications, you'll need to replace the piston.

Once everything has checked out and the ring gap has been established, it's time to install the new rings on the piston using the ring spreader to prevent overexpansion or distortion. If you're not using a ring spreader, carefully spread them by hand and slip them into the ring grooves starting with the lowest ring (the oil ring), ending with the top ring (the compression ring). Be sure to stagger the piston ring end gaps around the piston for maximum sealing. Note that on some models, the third compression ring from the top of the piston has a tapered face and must be assembled with the stamped word *top* toward the top of the piston.

Now, it's time to reinstall the pistons back into the cylinder. Remember, all components—especially the pistons—should be reinstalled in their original positions if they are being reused. Also, you'll need to note in your service manual whether the pistons need to be installed in a certain direction. Some have a notch on the crown that has to be oriented toward the front of

Before installing new rings on a piston, squarely position each ring in its respective cylinder, measure the end gap as shown, and compare it to the engine specs.

Insert the base of the piston into the bore and compress the rings with an appropriate-sized ring compressor. As an alternative, you may wish to put the compressor on the piston ahead of time, leaving enough of the base exposed to slip it into the cylinder or sleeve.

the tractor. Finally, lubricate the pistons and cylinder walls with clean engine oil.

Using a suitable ring compressor to compress the piston rings, place the piston into the bore. Basically, a ring compressor is a sleeve that fits around the piston to compress the rings enough to allow the entire piston to be slipped into the bore. Be sure the ring compressor is perfectly clean on the inside, and gently tap the piston down into the cylinder. Be careful to ensure the connecting rod studs don't scratch the cylinder walls or the crankshaft journal as you're installing the pistons. As extra insurance, you can always place pieces of plastic or rubber tubing over the connecting rod studs during piston installation.

At this point, all that is really left is to apply a light coat of engine oil to the connecting rod bearing inserts and install the bearing caps on the connecting rods.

Before you reassemble the engine, you should slip a feeler gauge in between the camshaft and its bushings to see if they have exceeded their useful life.

Since the rod bearing caps are already off, it's a good idea to also perform measurements to see if the journals need adjustment or replacement. On many machines, adjustment will simply involve removal of one or more shims, but if you don't do it while you're replacing the rings, you may have to go through the whole process again in the near future. For more specifics on determining serviceability of these components, refer to the following section.

Engine Block Preparation

Before you do anything else with the block you'll need to check it for hairline cracks. Depending upon the size of any cracks you find, you have a couple choices. One is to find a new block at the salvage yard. The other is to have the block welded by a competent welder. Chris Pratt notes, "Although there are probably weak points on several machines, a common example is the lower right corner on a Farmall Cub engine. This flange commonly cracks and will be a persistent oil leak once the engine is assembled."

Another more common problem, he says, is the lack of flatness of the mating surface between the head and block. To check both surfaces, you'll need a straight edge, feeler gauges, and your tractor shop manual, which will provide the allowable tolerance. On sleeved engines, which is the case with all but the Cub models, this tolerance takes on an extra importance because the sleeve stand-up (or how far the sleeves stick out of the block) must be taken into account. If there are radical differences between the cylinders following assembly, you'll need to locate the problems. These can include dirt under the sleeve flange, the sleeve setting down where the O-rings fit, or distortion of the lower mating surface. This can cause leaking at the head gasket, seepage of oil into the coolant at the base of the sleeve, and distortion of the sleeve that hampers free movement of the piston.

Main and Rod Bearings

The main and rod bearings can generally be lumped together, since the methods used to measure them are identical. The measurement you are looking for here is the existing size of the crankshaft pin or journal. Using this figure, you can calculate which undersized replacements will bring the crank or rod journal tolerance back to factory specification.

The first step in measuring the crankshaft and bearings is to check the surfaces for scoring. If scoring is minimal, it can generally be remedied by having a machine shop turn the crank on a lathe. However, this will cause the journal to be undersized to its original specification. If the scoring or damage is too bad, you'll simply have to look for a replacement crankshaft.

Assuming the crankshaft passes the initial inspection, use a caliper to determine if it has equal wear across the surface. If the journal has more wear on one side than the other, you will again have to look at having it turned or replaced. Finally, check to see if it has worn unevenly around the diameter of the journal, creating an oblong cross section. This can be done by taking a measurement, rotating the caliper around a quarter turn, and taking another measurement. Repeat this process a few times and you will quickly get an idea whether it is serviceable. The better manufacturers' manuals will even explain what are acceptable tolerances in this respect.

If the crankshaft does exceed the manufacturer's wear tolerances, you have two options. The first is to purchase a reground crankshaft. While this is expensive, it has the benefit of simplifying your measurements, since new bearings that have been pre-measured for the shaft are nearly always provided with the replacement. In this case, you are done with this job.

The less-expensive alternative is to have the crankshaft ground by a local machine shop. While the price is generally reasonable, there is often a wait, especially during certain times of the year, since many of these shops also work on specialty engines, such as those used for racing or tractor pulling.

Before you take a crankshaft in to be ground, though, be sure you can get the right-sized bearings to cover the amount of material the machinist will remove before you have the shaft ground. Otherwise, you may have wasted your money on the machining.

The other journal- and bearing-measurement process is accomplished by putting the whole assembly back together with Plastigage inserted between the bearing shell and the crank journal. Some manuals will talk about using shim stock to measure the clearance. However, since the advent of Plastigage, this technique is seldom, if ever, used anymore.

You should be able to find Plastigage at any good auto parts store. You'll find that it comes in different colors, like red, green, and blue. The color is a universal code for the range of clearance each particular plastic thread is capable of measuring. For example, red Plastigage is designed to measure a bearing clearance of 0.001 to 0.004 inches. If you take along your service manual, or tell the parts person how you want to use it and how much clearance you need, you shouldn't have to worry about colors. Any knowledgeable parts salesperson should be able to help you find the right size.

Basically, Plastigage is an impregnated string that squishes flat as it is squeezed between the journal and bearing shell. To use it correctly, cut a strip of Plastigage wide enough to go across the journal; then reassemble the shell and cap with the strip placed between the journal and shell; and torque the bolts to the proper specifications. Never turn the crank during this process.

Next, remove the bearing shell and compare the width of the Plastigage against a scale on the package to determine your exact clearance. Once you find this value, you can determine how much oversize will be required to bring the clearance back to that required by your manual.

As you've probably noticed, you have now measured the journals twice, once with a caliper and once with Plastigage. But with many old tractors, this is important. First, the caliper finds the irregularities and gross undersizes that necessitate crank welding, grinding, or total replacement. The Plastigage process is needed to determine whether shimming is required during reassembly. If Plastigage is the only measurement you use, it may be hard to spot irregularities like conical or oblong journals. On the other hand, you virtually can't measure for the proper clearance without Plastigage.

Now that we've gone through all of that, it's worth mentioning that there are a few exceptions. The 10-20, F-20, F-30, and W-30 are the most notable, as they use two ball bearings to support the crankshaft. Both are mounted in cages bolted to the crankcase.

According to the service manual, the condition of the bearings can be determined by removing the oil pan and shaking the crankshaft up and down for radial clearance and forward and back for end clearance. Since the bearings are non-adjustable, they must be replaced if the clearance becomes excessive. This, of course, means removing the engine from the tractor and removing the flywheel in order to reach the rear bearing.

Push Rods

Push rods can bend if the valve timing was off or even if a valve was adjusted to be open all the time. Visibly bent push rods should be replaced and all others should be checked. Checking them can be done with a perfectly flat surface and feeler gauges.

Although some people claim they can straighten push rods, it's usually best if you replace them, even if it means going to a salvage yard for better ones.

Camshafts

Although camshafts can bend, it's not likely, since the push rods tend to sacrifice themselves much sooner. The more common problem with the camshaft is worn or scored bushings. This can result when the surface went too long without oil or if a foreign object lodged in between the bushings and camshaft.

While it is possible to get the shaft turned, you'll need to first check to see if oversize bearings are available, since only standard-size bearings are normally sold for old tractors. If no oversize bearings are available, you can still have a machine shop make new bushings. Another alternative is to purchase a used or reconditioned camshaft.

Among his spare parts, Paul Cummings managed to acquire a complete crankshaft with pistons for a Model C. It even has the bushings taped to the crankshaft.

Due to some scoring on the shaft, this crank would still need to be turned before it could be reused.

Jeff Gravert, a full-time tractor restorer from Central City, Nebraska, measures the crankshaft journal to check for out-of-round, taper, and wear.

Connecting rod and main bearing clearance can be easily checked for tolerance using the appropriate size of Plastigage.

Rocker Arms

Like the push rods, you'll need to check the rocker arms for straightness and smooth profile. Rocker arms can be distorted, which can not only make adjustment difficult, but cause the push rod to slip when combined with a slightly bent push rod. On certain models, you'll also need to make sure the rocker arms are getting adequate lubrication. For example, on the F-20, F-30, W-30, and 10-20, the valve assembly is lubricated from a trough containing packing, which is mounted above the lever assembly. This trough must be manually filled with oil through an opening in the valve cover. For this reason, it's important that the trough be reinstalled with the three 3/16-inch holes facing toward the valve side of the engine.

Similarly, lubrication to the valve lever assembly on the W-40 is provided by a hollow cylinder head stud (third from rear on left hand side). Spurt holes in the exhaust valve levers then distribute oil to an oiler pad mounted above the lever assembly. Hence, correct reassembly is vital.

Valves and Valve Seats

Faulty valve action is one of the main reasons for loss of power in an engine. Although carbon, corrosion, wear, and misalignment are inevitable products of normal engine operation, the problems can be minimized with high-quality fuel and valve tune-ups.

Carbon is a byproduct of combustion, so it's always going to cause some problems, like fouling spark plugs, which makes the engine miss and wastes fuel. But valve seats can also become pitted or be held open by carbon particles. Carbon deposits can also insulate parts and cause them to retain heat, compared to clean metal, which tends to dissipate engine heat. This increases combustion-chamber temperature and causes warping and burning.

Finally, unburned carbon residue can gum valve stems and cause them to stick in the guides. Deposits of hard carbon, with sharp points on them, can even become white hot and cause pre-ignition and pinging.

Consequently, valves need to be carefully checked for warping, burning, pitting, and out-of-round wear, especially on the exhaust valve, since it is exposed to the high temperatures of exhaust gases.

Burning and pitting are often caused by the valve failing to seat tightly due to carbon deposits on the valve seat. This in turn permits exhaust blow-by, although it may also be due to weak valve springs, insufficient tappet clearance, warpage, and misalignment. Warpage occurs chiefly in the upper valve stem due to its exposure to intense heat.

Out-of-round wear follows when the seat is pounded by a valve whose head is not in line with the stem and guide. Oil and air are sucked past worn intake valve stems and guides into the combustion chamber, causing excessive oil consumption, forming carbon and diluting carburized fuel.

Misalignment is a product of wear, warpage, and distortion. Such wear, which is often hastened by insufficient or improper lubrication, will eventually create sloppy clearances and misalignment. Distortion, on the other hand, is generally caused by unequal tightening of cylinder head bolts.

Valve Guides

Valve guides tend to warp because of the variation in temperatures over their length. Consider, for example, that the lower part of the guide is near combustion heat, while the upper is cooled by water jackets.

Any wear, warpage, or distortion affecting the valve guides destroys its function of keeping the valve head concentric with its seat, which obviously prevents sealing.

Valve Springs

Valve springs must be of a uniform length to be serviceable. To check the springs, place them on a flat, level surface and use a straight-edge ruler to determine whether there is any irregularity in height. Unequal or cocked valve springs should be replaced.

Spring tension that is too weak will also allow the valves to flutter. This aggravates wear on the valve and seat, and can result in valve breakage. If the springs are less than 1/16 inch shorter when compared with a new one, they should be replaced.

Pistons and Sleeves

Just as you did with the crankshaft, you'll want to examine the piston and sleeves for any visual damage, such as scoring, out-of-roundness, and greater width on the sleeve at the highest point of piston travel (not including the ridge that forms above the high point of travel). Scoring will indicate the need for replacement regardless of the measurements. The other conditions can be determined through precise measurement.

To check for out-of-roundness, use an inside micrometer to take a measurement at the top and bottom of the cylinder. Then repeat the process at right angles to the original measurements. With these numbers, you can see how conical the cylinder is and how oblong it has become. Repeat the process using your caliper on the piston. There will usually be a factory specification for what is acceptable.

Some pistons are cam ground, meaning they are supposed to be slightly out of round. Similarly, there is generally more wear at the top of the bore than at the bottom. However, the amount of deviation from the factory specifications on both the piston and the bore will determine whether the parts need to be replaced. If the engine is the type that doesn't use a sleeve, which is case with the Cub, the only option will be to have the block bored out and a sleeve installed in the cylinder bore.

If the differences at the top and bottom are acceptable, you may still need to determine if there is too much distance between the piston and the sleeve/bore. This will be most noticeable at the top of the stroke. Most shop manuals suggest using a long feeler gauge of a certain size placed between the piston and sleeve with the piston inserted in the sleeve. A scale is then used to pull the feeler gauge out of the bore. The amount of drag required to pull it free, as measured on the spring scale in pounds of pull, determines its serviceability. This process needs to be done without the rings in place, and the measurement is usually taken 90 degrees from the piston pin hole.

There is an additional series of sleeve measurements still to be made. You may be wondering why these measurements are needed beyond those you have already taken. Well, the real goal is to avoid any unpleasant experiences when purchasing new pistons and sleeves, like having to send them back and try again. The reason this happens is that many tractors had more than one sleeve size for a given model. A few tractors even had different sizes embedded in the middle of their serial number ranges.

To make matters worse, it can take some serious research to ensure that your engine was not replaced with one that is two years older and, thus, has a different block. The helpful measurements here are the bore and stroke. This is the distance across the piston and the distance the piston travels. You should also measure the thickness of the sleeve. Lastly, a few engines can be identified by measuring the outside diameter of the sleeve or the outside diameter of the base of the sleeve where it seals to the block.

According to Pratt, tractors that should always be checked for these types of differences are the Farmall A, B, C, Super A, and Super C. "These are ones that I have commonly had the serious problems with, but there are probably many more," he says. "In some cases, the sleeves will fit, but you may end up with a smaller cubic-inch displacement and less power when your overhaul is complete."

Piston Pins

The piston pin is difficult to measure without using calipers to measure the pin—and the results of this may be meaningless to the home mechanic anyway. If your tractor does not use bushed piston pins, then bearing replacement may be inexpensive and quick. If it does use bushings, then refer to the service manual for indications of wear and acceptable tolerance. The main objective is to ensure there is no detectable looseness. Some models call for "hand push" tightness, while others call for a "light press" fit.

Since this piston was frozen in the sleeve, it's obvious the rings are also frozen into the ring grooves. Should it need to be replaced, pistons on most models are not available as individual parts; they're only available as matched units with the sleeves.

Structurally, this piston rod appears to be in good shape for reuse. All that was required was a new set of rod bearings.

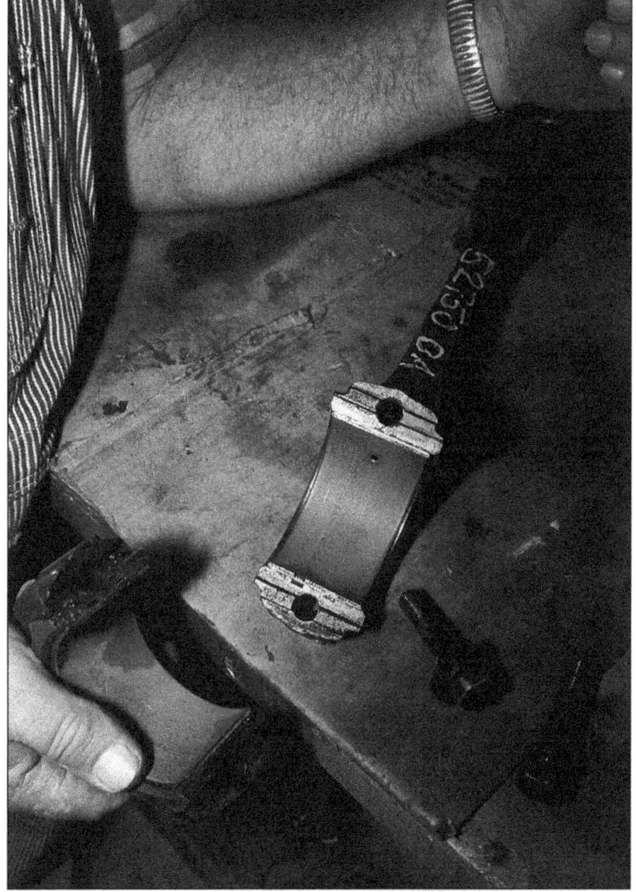

Most connecting rod bearings are of the slip-in, precision type, and are available in both standard and various undersized sets.

Should it become necessary to pull the sleeves for replacement or an engine rebuild, you'll need a sleeve puller.

Farmall enthusiast Paul Cummings shows how the flange on the sleeve puller engages the bottom of the sleeve for removal.

New sleeves should be carefully unwrapped and examined prior to installation in the block.

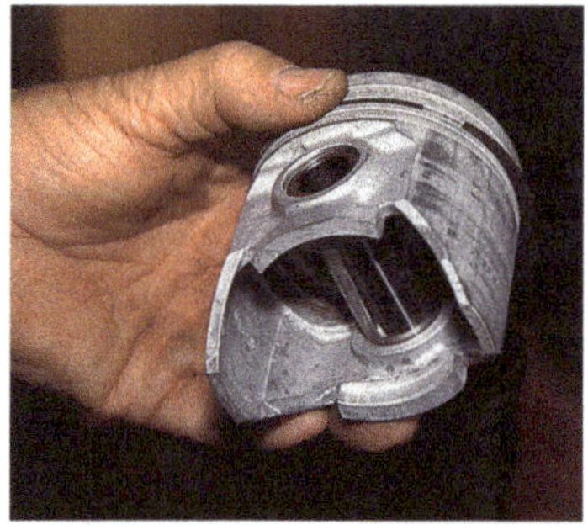

Each new sleeve in a rebuild kit comes with a matching piston and rings.

This sleeve with piston is designed for a Farmall F-20

Rebuilding the Engine

If you're planning to completely overhaul the engine, you may want to consider purchasing an engine rebuild kit—assuming there is one available for your tractor. Most have all the replacement parts you'll need for overhauling an engine without going to the store to purchase separately gaskets, special measuring tools, or miscellaneous parts. But don't let the matching sleeves, pistons, rings, and bearings lead you into a false sense of security. They, too, need to be measured and installed in a workshop environment.

If you are going for a total rebuild, your best bet at this point is to have the engine block, head, and any other major grease-carrying components professionally cleaned in a hot chemical bath. This process will not only strip those components of grease and paint, but it will also remove carbon, varnish, and oil sludge from both the inside and outside of each part.

While the engine components are out for cleaning, it's also a good idea to have a professional magna-flux the engine block and head for cracks that are not visible to the naked eye and, as a result, would have been overlooked in your original inspection. This investment alone could save you a lot of time and money.

Meanwhile, smaller parts can be cleaned in the shop using a solvent tank, then dried, inspected for integrity, and placed on a clean rag. When the engine comes back from the machine shop, it should ideally be mounted on a rotating engine stand and prepared for reassembly.

Use an air hose from your air compressor to blow out the engine block. Be sure to hit all the openings and bolt holes to remove any residue that may be left. A clean solvent rag can be used to wipe off the internal surfaces to remove any film left over from the tanking process. Hopefully, the machine shop has reinstalled the casting plugs (sometimes called *freeze plugs*) and oil galley plugs.

It is generally recommended that the head be reworked in its entirety by the machine shop, as well. Pressing in valve guides, getting the correct angles

Don't forget to reinstall the oil trough and packing on Regular, F-20, F-30, and W30 models. Mounted above the lever assembly, the trough is manually filled with oil through an opening provided in the valve cover.

on the valve seats, setting the proper valve-to-head re-cession, and measuring and milling warped surfaces, among other things, are really beyond the scope of the average home shop level.

Crankshaft and camshaft measuring, grinding, and polishing are also out of the realm of the home shop. It is wise, though, to always double-check the machine shop's work with a feeler gauge, dial indicator, and Plastigage. It is important to check the crank end play with a feeler gauge and also to check each bearing journal with Plastigage before starting the reassembly procedure.

If your rebuild includes cylinder sleeves, you may also want to look at having the machine shop install these, as well. Just remember that the pistons and sleeves come as matched sets; so you'll need to identify which piston goes in which sleeve before turning loose of half the kit.

If you're going to be putting in a set of dry cylinder sleeves yourself, one restoration specialist recommends putting the sleeves in the freezer for a couple of hours. This will cause the metal to shrink just enough to slip into the cylinder bores a little easier. Place the block on a hydraulic press and line it and the ram up with the sleeve. Place some cylinder sealer around the upper end of the sleeve, install an adapter on the top of the sleeve that will mate to the ram, and quickly but gently press the sleeve into the bore.

Don't stop midway for any reason as the natural heat from the block will quickly cause the sleeve to expand, increasing the risk of breakage. Only stop when the sleeve is flush with the top of the block.

Wet sleeves, on the other hand, should fit into the crankcase bores full depth and be free to rotate by hand when tried in bores without sealing rings. After making a trial installation without sealing rings, remove the sleeves and install new sealing rings into the grooves in the crankcase. Now, wet the end of the sleeve with a thick soap solution or equivalent and install. If the sealing ring is in place and not pinched, it should take very little hand pressure to press the sleeve completely into place. To test the sleeve integrity, fill the crankcase (cylinder block with cold water and check for leaks around the sealing ring grooves.

If you're simply honing the old sleeves or the cylinder walls, you can refer back to the piston ring replacement section for this process. Cylinder sleeves that come in a kit with the pistons do not require honing after installation, but should be checked for distortion or high spots.

Now that the sleeves are in place or the cylinders have been honed, the next step is to match the pistons to each cylinder. Refer to your shop manual and locate the piston-to-cylinder wall measurement. Generally, it should be around a 0.004–0.007-inch sidewall clearance. To obtain this measurement, you'll need to locate a long feeler gauge, commonly called a ribbon gauge. Install the piston in the bore and see if the specified feeler gauge will slip in next to the piston. The proper fit is when the ribbon gauges can be pulled from between the piston and sleeve with a specified inch-pound pull as verified by an inch-pound scale. If you don't have the recommended scales, you should at least make sure the feeler can be pulled out freely with moderate pressure and without binding. You should also check the piston at the top and at the bottom of the cylinder for proper clearance in case the sleeve is tapered. Don't assume that if the piston freely falls through the bore that the clearance is adequate. If the clearance is inadequate, the cylinder bore (or sleeve) will have to be honed and then rechecked. Take the time to do it right.

Each piston will come with a certain number of piston rings. Unwrap the rings and lay them next to each piston in the order of installation. Basically, the instructions for measuring and installing the piston rings are the same as for piston ring replacement, so refer to the previous section in this chapter about piston ring replacement.

Before reinstalling the cylinder head, slip a new head gasket over the mounting bolts, making sure it is oriented correctly.

Every thorough engine overhaul should include an inspection of the oil pump, since it provides the life blood to engine parts.

Having been cleaned, inspected, and primed, this spare engine block is ready for sleeves and pistons in the event it's needed.

Adding the valve cover should finish up this Model C engine overhaul.

When installing new bearing shells, make sure the rod and bearing cap numbers are in register and the shell has been well lubricated.

Once the piston rings have been installed, the next step in assembly is to check the piston pin-to-bushing tolerances. Once again, this is a close tolerance and is best done by a competent machine shop. An insertion pressure of pin to bushing will be slightly looser than that of the pin to piston. When dealing with clearances of 0.0002 inch or less, it is important to have the proper measuring tools. A pin too tight or too loose will cause a piston and rod to break apart under the stresses of engine operation.

Valvetrain Overhaul

Depending upon the condition of the engine when you started, you may or may not have to do major work on the valvetrain. If the engine was running when you acquired the tractor, a good visual inspection and adjustment of the tappets may be sufficient.

However, when water gets into an old engine, it seems that the valves are among the first components to suffer. As a restorer who likes to focus on orphan tractors, Bill Anderson of Superior, Nebraska, has seen his share of valvetrains that not only required one or more new valves, but valve seat restoration, as well. Generally, this can be accomplished by using a valve seat tool to regrind the seats. However, if damage to one or more valve seats is severe, it may be necessary to have a machinist cut recesses into the block or head that will accept special hardened valve seats.

To inspect, grind, or lap the valves and seats, you'll first need to remove them from the head or block, depending on whether it is an overhead-valve engine or a valve-in-block engine. To do this, you'll need a valve-spring compressor so you can compress each of the valve springs. Once the spring is compressed, you can remove the clip or keeper from the base of the valve spring. At this point, the valve can be lifted out. Be sure to keep the valves separate so each one can be replaced in the original seat.

Depending upon the condition of the valves, you'll have one or more options. If the valves have notches or burned sections, you'll need to replace them. However, if the valve faces and seats are simply a little rough and dirty, they can be cleaned and renewed with valve lapping. To do this, make sure the valves are returned to their original position after inspection, so you're lapping each valve into its original seat.

Before you install the valve, though, place a small bead of medium-grit valve-lapping compound around the face of the valve where it mates to the valve seat. Now, using a valve-lapping tool, rotate the valve back and forth in the seat, alternating between clockwise and counterclockwise rotation. In the simplest form, a valve-lapping tool is little more than a suction cup on a stick that you spin in your hands. Other, more sophisticated versions have a suction cup on a shaft driven by a device that looks a little like an egg beater. Some older-style lapping tools, however, engaged a slot or recesses in the valve head.

At any rate, you can move on when you have a smooth, clean surface all the way around the valve face with a matching surface on the valve seat. If necessary, add more compound during the process.

Valve lapping should not be a substitute, though, for having the valves professionally reground. If they're in rough enough condition such that a few minutes of lapping doesn't polish up the mating surfaces, you're probably better off having a machinist grind the valves and seats. This basically consists of clamping the valve stem into what looks like a large drill chuck and turning them against a grinding wheel that renews the face to the correct angle.

All that is left now is to thoroughly clean all the parts, including the keepers. Then, check the valve springs to make sure they meet service manual specifications and reassemble the valvetrain, coating all parts with clean engine oil as you go.

Once the valves and springs have been assembled back into the head or block, you'll need to finish up by adjusting tappet clearance according to specifications. Correct clearance contributes to quiet engine operation and long valve-seat life. Insufficient clearance causes the valve to ride open, resulting in lost compression and burning. Too much clearance retards timing and shortens valve life.

Oil Pump Restoration

Any thorough engine overhaul should include an inspection of the oil pump. Although oil pumps can be of the vane type or gear type, almost all vintage Farmall tractors use the latter. Like the gear pumps used in some hydraulic systems, the pump uses two gears that mesh together to create an area of high pressure on one side of the gears and low pressure on the other side. As one gear drives the other and the teeth mesh, oil is carried around the outside between the gear teeth and housing. Inspection involves cleaning or replacing the pickup screen and checking the gear surfaces and housing for wear and cracking. Be sure to also check for worn bushings or bearings and ensure that all oil passages are clear. Check your repair manual if you have any doubt about wear criteria.

Finally, make sure you use the correct thickness of gasket between the pickup screen and gear housing when you put everything back together. On a number of models, including the Farmall A, B, C, H, and M, gaskets between the pump cover and body can be varied to obtain the recommended gear end play during reassembly. Ideally, the pump should also be primed before replacing the oil pan. One way to do this is to pack it with a lithium-based grease, commonly called white lube, before replacing the oil pan.

A valve spring compressor is used to compress the valve springs for access to the keeper, which is a small pin, clip, or wedge that needs to be removed to remove the spring.

Once they've been removed, carefully check each valve for warping, burning, pitting, rust damage, and out-of-round wear. It's also important to check the valve springs for irregularity in height or spring tension.

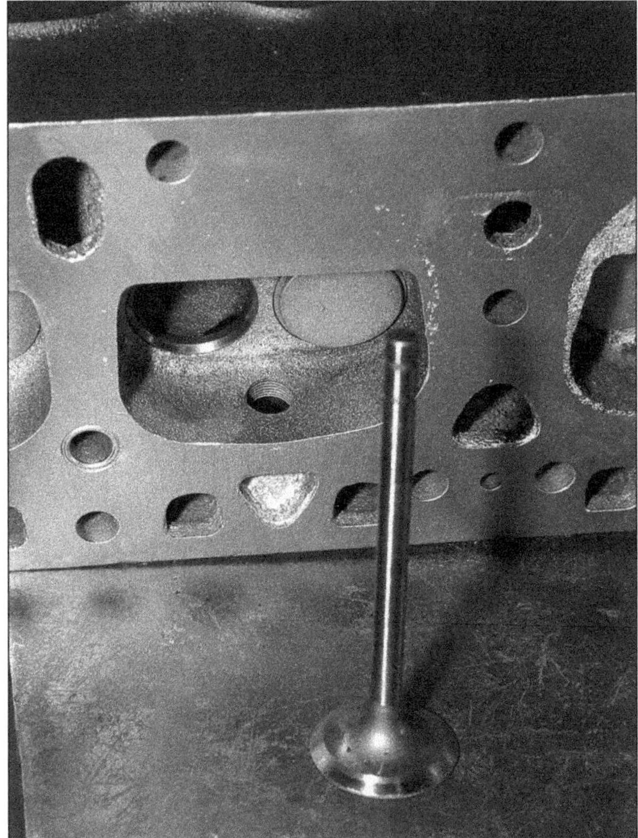

When properly finished, both the valve and the valve seat will have the type of smooth surface that ensures a tight seal.

Valves that are slightly pitted or worn may need to have the face ground on a valve grinding machine.

Using the valve lapping tool, which usually has a suction cup on one end, rotate the valve until the face and seat have matching polished areas that are smooth and clean.

The last step in valve train overhaul is reinstalling the valve springs and adjusting the tappet clearance to specifications.

Chapter 7

Clutch, Transmission, and PTO

When it comes to transmissions, it's usually a good news, bad news situation. The good news is that transmissions were probably the toughest-built items on antique tractors. In fact, about the only time a tractor restorer faces a big problem is when a transmission failure was the reason the tractor was permanently parked in the tree row in the first place—and that's the bad news. As an example, some tractor models, including the Farmall M and H, have little clearance between certain gears and the bottom of the transmission case. In the event of bearing failure, ball bearings were known to occasionally drop to the bottom of the case. Now, picture the gear teeth catching that ball bearing, but having insufficient clearance to move it out of the way. The only place for the little steel ball to go was through the bottom of the transmission case.

If the tractor is in running condition, you should have had the opportunity to drive it and run it through the gears before you made the purchase. Hence, you should have an idea where the trouble lies. Often it will be with one of the gears that saw a lot of field work, particularly if there were only three or four gears in the transmission. As was stated in the Troubleshooting section, this was often second gear, which becomes obvious when you hear the whine when you test drive the tractor.

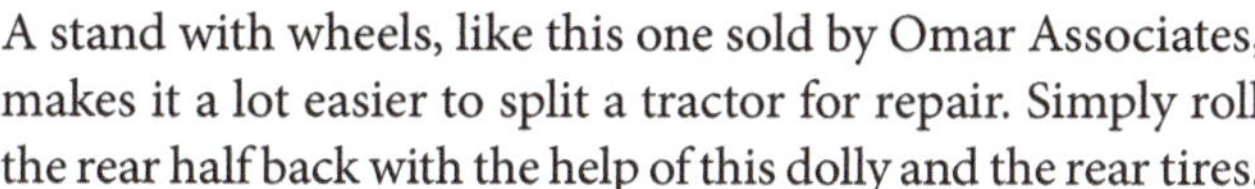

A stand with wheels, like this one sold by Omar Associates, makes it a lot easier to split a tractor for repair. Simply roll the rear half back with the help of this dolly and the rear tires.

Quite often, you'll run across an F Series model that has been fitted with an aftermarket road gear, such as this Behlen model.

Other things to watch out for are the effects of dirt and water. Just as with a modern-day transmission, dirt is the gearbox's biggest enemy. If the shift boot has worn or rotted away, then there's probably water in the transmission, too.

Still, there's a good chance that you will get lucky and find that all it takes to get the transmission in working order is to drain it, clean it up, and refill it with fresh transmission fluid. In fact, most tractor enthusiasts say that's all that has been needed on the majority of the tractors they have restored.

That's probably not the case with the clutch, though. If during your test drive—assuming the tractor is in running condition—you found the clutch slipping or chattering, you'll need to take a closer look. Most likely the problem is a worn clutch plate. At any rate, the clutch is probably going to need more attention than the transmission.

Transmission Inspection and Repair

As was stated earlier, transmissions used in tractors built in the 1920s, 1930s, and beyond were generally heavy-duty enough that they don't present any major problems, even after sitting for several years. Still, it is a good idea to at least open up the transmission to check out the gears and clean them up.

Several restorers say they like to drain any fluid they find in the transmission case and replace it with diesel fuel. Keep in mind that after sitting for several years, the fluid can be about as thick as tar. While you're draining the transmission, keep an eye out for pieces of metal or fine shavings that can tip you off to problems. Many times, there will be a magnet built into the drain plug to help catch these things.

Now, drive the tractor around the yard for several minutes to circulate the diesel around the transmission case, coating and rinsing all the gears. If the tractor can't be driven but can be towed, that would be your next-best option.

If you can't use gear action to do the work, though, you'll need to clean every part you can reach with cleaner and a stiff brush.

Next, inspect all the bearings, shafts, and seals for damage. Rotate the gears with your fingers as you go through the cleaning and inspection process. At the same time, check for looseness and rough action. If it wobbles or shakes, it's probably going to need repair.

Keep in mind, too, that transmissions built in the early part of the twentieth century were not synchronized like they are today. So if farmers didn't wait until the tractor stopped before shifting, they had a tendency to round off the gears. This was particularly the case with the "road gear" in many late 1930s and 1940s tractors. Consequently, you may find a gear or two that needs replacement due to ground teeth.

Unfortunately, the biggest challenge with transmission repair may not be the removal and replacement of a defective gear or bearing, but finding the necessary part in the first place—or finding it at a reasonable price. One option is to check at swap meets and salvage yards. Armed with all the measurements, you might also be able to find similar gears from another model. Another option, if you're faced with a transmission that needs extensive repair is to buy a rebuilt transmission or a used transmission that's in better shape from a salvage yard.

While inspecting or rebuilding the transmission, it's also important to pay particular attention to all transmission bearings. It should go without saying that any bearings and seals that look bad should be replaced.

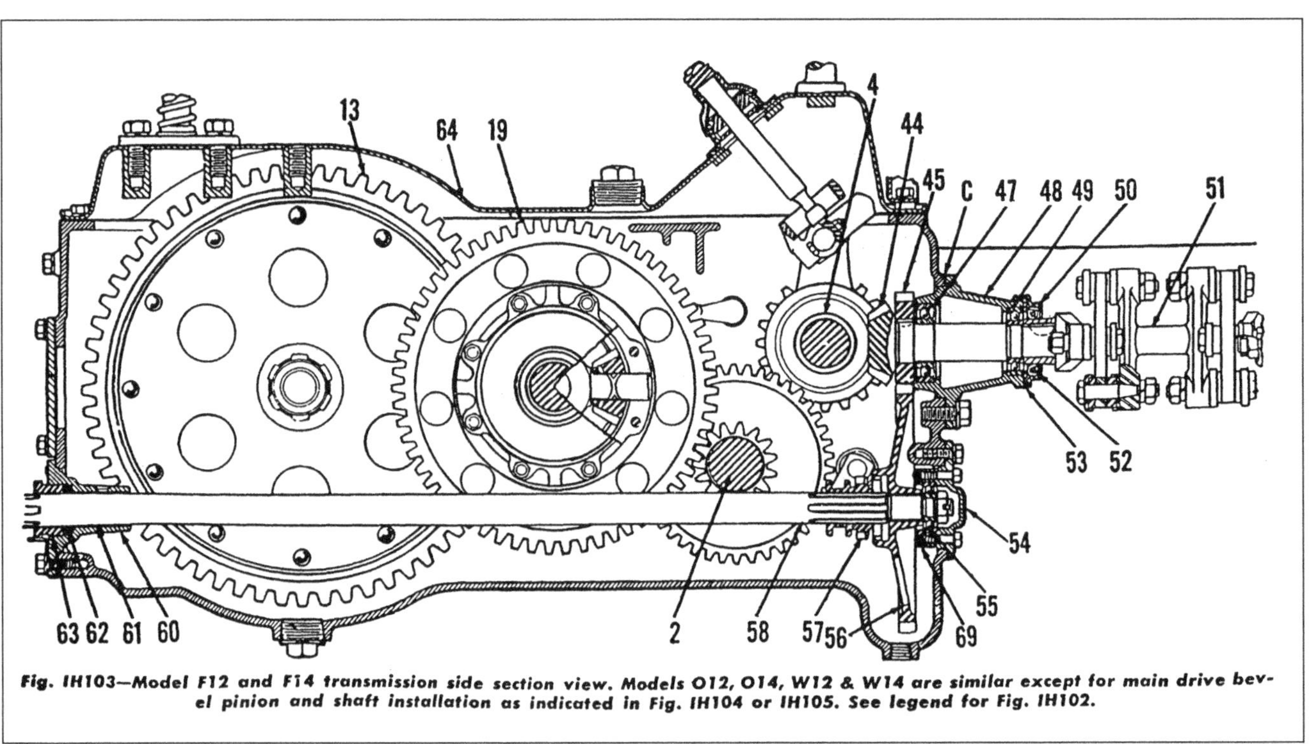

Fig. IH103—Model F12 and F14 transmission side section view. Models O12, O14, W12 & W14 are similar except for main drive bevel pinion and shaft installation as indicated in Fig. IH104 or IH105. See legend for Fig. IH102.

This side-section view of the F-12 and F-14 transmission shows the simplicity of the early three-speed gearbox and final drive. The O-12, O-14, W-12, and W-14 have a similar configuration. (Courtesy of Jensales)

Torque Amplifiers

While it's true that most transmissions used on vintage tractors were limited to only four or five gears, International Harvester was one of the true innovators when it introduced the Torque Amplifier as early as the Model M and W6. Models equipped with the Torque Amplifier were appropriately designated the MTA and W6TA. Later models, such as the 460 and 560, included a Torque Amplifier as standard equipment.

Basically, torque amplification is provided by a planetary gear reduction unit located between the engine clutch and transmission. The unit is controlled by a hand-operated, single-plate, spring-loaded clutch. When the clutch is engaged, engine power is delivered to both the primary sun gear and a planet carrier. This causes the planet carrier to rotate as a unit and the system is in direct drive. When the clutch is disengaged, engine power is transmitted through the primary sun gear to a larger portion of the compound planet gears. As a result, an overall gear reduction of 1.482:1 is obtained when the torque amplification clutch is disengaged. Best of all, the reduction could be obtained without clutching and without interrupting engine power.

Due to the complexity of these types of units, it is often beyond the capability of most restorers to make repairs. Repairing the torque amplifier also means removing the entire clutch housing, not just splitting the tractor, since the torque amplifier unit is located at the rear of the housing between the engine clutch and the transmission. Consequently, you may want to have the transmission serviced by the local equipment dealer or a qualified mechanic. If you choose to attempt it yourself, make sure you're equipped with a detailed service manual.

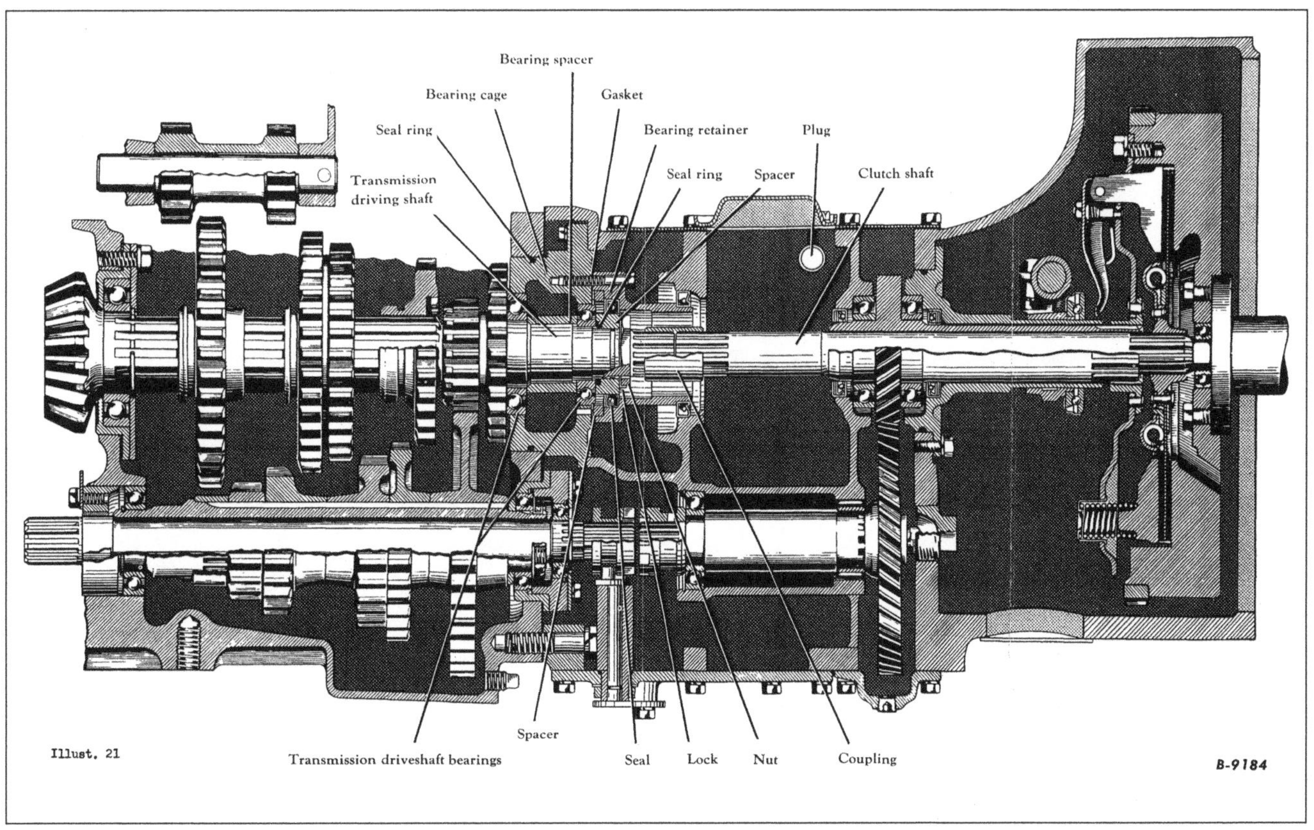

This cross-section drawing shows the Farmall 300 transmission and clutch housing for a tractor without the Torque Amplifier. (Courtesy of Jensales)

Since the transmission is still sealed and there are no missing gear shift boots, there's a good chance the transmission on this tractor will need little work.

Thanks to a jointed clutch and transmission coupling on F Series models, it's possible to overhaul the clutch and gear set without splitting the tractor. Otherwise, be sure to check the seals on each end and make sure the coupling doesn't have any free play.

The first step in transmission overhaul is cleaning up the gears and shifting mechanism, and inspecting the gears for missing teeth.

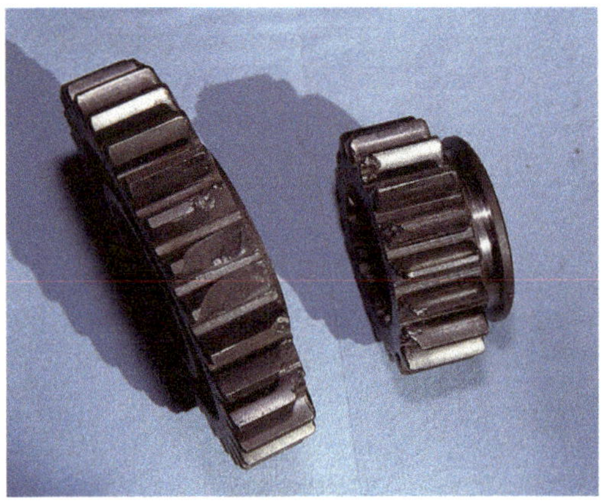

It's easy to spot the damage on these gears. The chipped teeth and rounded corners were caused by "speed shifting" a transmission before the tractor came to a stop.

In most cases, the belt pulley, whether standard or optional, is driven off the top of the transmission case.

The belt pulley assembly should be checked for gear wear and bearing integrity, just as you would with transmission and PTO components.

In order to service or repair the Torque Amplifier on models so equipped, it's necessary to remove the entire clutch housing/torque tube.

Clutch Inspection and Rebuilding

Once you've finished with the transmission—either having cleaned it and verified its condition, or replaced the appropriate gears and bearings—it's time to move on to the clutch. Essentially, these two components work together anyway.

Overhauling a clutch is never an easy job. But it's one of the most important, partly due to the safety issues. Obviously, you want the tractor to stop when you push in on the clutch and apply the brake. But if you have a tractor on which the clutch has rusted to the shaft, it's also possible for the tractor to move when the engine is started, even if the transmission is in neutral.

While some of the early tractors, including steam tractors, had positive-engagement clutches, the move toward internal-combustion engines necessitated the use of a friction clutch that could be slipped as the drive was engaged. As is the case with most vintage tractors, all Farmall models use a plate clutch that is attached to the engine flywheel. This means the tractor needs to be split at the bell housing or the engine removed for clutch replacement or inspection.

If you drove the tractor before you purchased it or started restoration, you should already have an idea whether the clutch needs attention. Usually the trouble is obvious and falls into one of three categories:

- The clutch slips, chatters or grabs when engaged
- It spins or drags when disengaged
- You experience clutch noises or clutch pedal pulsations

If the clutch slips, chatters, or grabs when engaged, or drags when disengaged, the first thing you should do before tearing into the clutch itself is see if the clutch linkage is improperly adjusted. In some cases, free travel of the clutch pedal is the only adjustment necessary for proper operation of the clutch—assuming your tractor is equipped with a foot clutch. Free travel is the distance the clutch pedal can be pushed before resistance is met. Refer to the repair manual for your tractor for this dimension.

On tractors with a hand clutch, there should be a definite feel of over-center action when pushing forward on the hand lever. A slight pressure should be felt in the hand lever, then a definite release of pressure as the clutch goes into engagement.

It's worth noting, too, that the clutch linkage on some tractors, such as the late model Cub and models equipped with a Torque Amplifier, can be adjusted externally at the pedal linkage, while the majority of others, including the A, B, and C, require you to make the adjustments through a hand hole cover in the clutch housing.

Beyond clutch linkage adjustment, the cause of most problems are generally found in the clutch housing where clutch components are either worn, damaged, or soaked with oil, which is causing the clutch facing to slip. Most often, the culprit is pressure springs that are weak or broken, or the friction disc facings are worn. Hence, clutch restoration generally consists of cleaning and checking all parts for wear, replacing bushings in the sleeve, if necessary, and replacing the clutch lining.

Although the clutch lining is attached with rivets on the majority of vintage tractors, some tractors use a drive plate to which the friction disc is bonded. If available, your best bet is to replace the entire clutch plate with a remanufactured part. If you're working with an older tractor, though, you may have to have a machine shop reline the existing plates by riveting new linings in place.

Prior to reassembly, Paul Cummings spread out the clutch components for his Farmall Regular to show all the components.

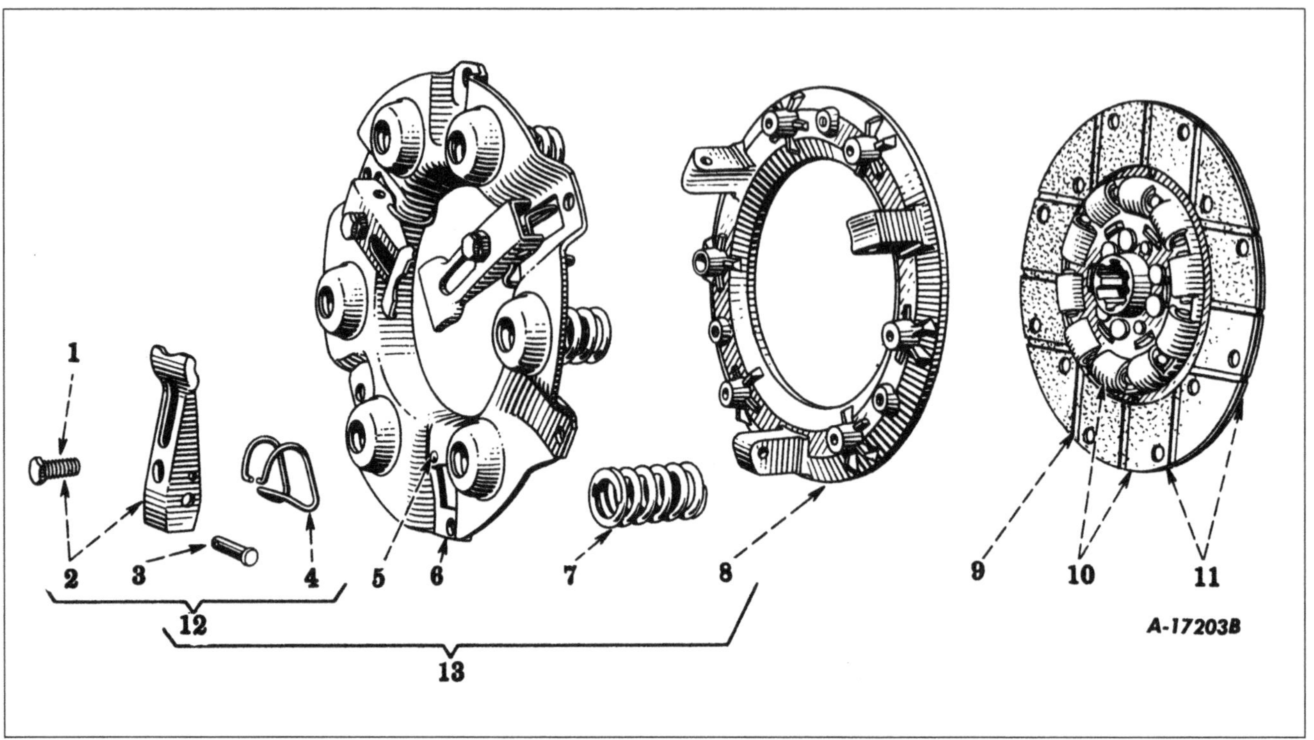

An exploded view in the parts manual can be as helpful as the repair manual when reassembling the clutch.

Clutch Throw-Out Bearing

Another potential problem spot is the throw-out bearing, which compresses the springs to release pressure on the plates when the clutch is pushed in. Characteristics associated with a defective throw-out bearing include squealing when the clutch is released and rough actuation. Replace the throw-out bearing if there is any hint of roughness, looseness, or discoloration.

Power Takeoff Repair

Although the first rear power takeoff appeared as early as 1918, the PTO never played a big role until a few decades later. Until rear-mounted PTO-driven implements began appearing in earnest, most farmers continued to use the belt pulley as their main power source.

Early PTO systems were also a pain to use with certain implements, since they were driven from the transmission. That meant the tractor had to be moving or in neutral with the clutch engaged for the PTO to operate. Many farmers can still remember mowing hay fields with a transmission-driven PTO. If the sicklebar started to plug, or you had to stop in the middle of the field, you could almost count on getting it plugged even worse, because the minute you stopped, so did the sickle.

Things changed, though, in the 1950s when International Harvester introduced tractors with a live PTO. The live, or independent, PTO utilized its own clutch within the transmission, which meant the PTO could continue to operate in relation to the engine speed rather than slow or stop its function as the tractor slowed or stopped. Within a few years, all tractors had a live PTO.

Whether your tractor is equipped with a transmission-drive PTO or an independent-drive system, you should check its condition as part of transmission inspection and repair. The most common problem tends to be seal leakage. Other ailments can include clutch problems with live PTO systems and worn gears and bearings. Due to the different variations used by tractor manufacturers, overhaul procedures are best explained in your tractor repair manual.

To access the clutch in most Farmall and IH tractors, you'll either need to pull the engine or split the tractor, particularly on models that use the engine and transmission as a structural support. One exception is the Model H with its 11-inch clutch.

The wearing and glazing on this clutch disc was reason enough for replacement.

The chip in this throwout bearing from a Farmall Cub makes the need for a replacement rather obvious.

The clutch on this Farmall Regular is pretty open and basic compared to newer models.

You can spot some of the wear, as evidenced by the scoring on the over-running clutch ramp on this Torque Amplifier.

The planet carrier, seals, and over-running clutch ramp were among the parts replaced during a Torque Amplifier overhaul.

Chapter 8

Final Drive and Brakes

Your first thought may be that the final drive and brakes have little in common and really don't go together in one chapter. But when you consider that the role of the brakes on a tractor is to stop one or both rear wheels, you can begin to see how the two fit together.

In general, there are two points of application with the brakes used on most vintage farm tractors. What's more, that braking point has a lot to do with the type of final drive you find on the tractor. For example, most models that used bull gears to drive the axles have brake housings located on the sides of the transmission/final drive case. Generally, these are expanding-drum brakes on a splined shaft that engage the bull gears or bull gear pinions, which is the case on most early Farmall tractors. So, in effect, the brakes aren't stopping the axles or the wheels, but rather the pinions or bull gears that drive the axles.

In contrast, tractors that utilize a pinion shaft and ring gear in combination with a spider gear set don't have a braking point on the gear sets themselves. Hence, the brakes are located on the drive axles or in the wheel hubs.

Differentials and Final Drives

Although there are a number of different types of final drives, you can count on all of them having one thing in common: They were generally built tough enough to take all the torque the engine and transmission could generate, and then some. The biggest exceptions, of course, were the Farmall 460 and 560 tractors, which introduced six-cylinder engines for more horsepower, while retaining the rear end that had been used since the M was introduced. As a result, the final drives of both tractors began to fail after about 300 hours of normal to heavy use.

International Harvester tried to solve the problem by revising the tractors on the assembly line and repairing tractors in the field. So, assuming your tractor is in running condition, the company fiasco shouldn't affect you unless you're planning on restoring a 460 or 560 as a work tractor. Most restorers look for earlier models anyway. The bottom line is that the final drive on most vintage tractors needs little attention other than replacing bearings and seals and changing the fluid.

Since the introduction of the first Farmall Regular, International Harvester has used a hypoid- or bevel-gear type differential to transfer engine torque from the transmission to the axles. Composed of a bevel pinion and shaft, bevel ring gear, and a set of pinion and side gears, the differential provides a means of turning the power flow 90 degrees and dividing the power between the two rear wheels. The differential also provides further gear reduction beyond the choices provided by the transmission for additional torque to the rear wheels.

Since the proper gear mesh is important, most tractor manufacturers recommend that if you replace the ring gear, you should also replace the bevel pinion gear, so you maintain a matched set.

Beyond the differential, certain Farmall models used various other final drive components. Some, like the majority of the letter and number series, take care of all gear reduction in the transmission and differential and connect the axle shaft directly to the wheel hub. Others, like the Regular, F-20, F-30, A, B, and Cub utilized another gear set at the wheel end of the axle. In addition to providing further gear reduction, which is the function of the bull gears on other models, the final drive gear set and housing raised the tractor for additional ground clearance. It wasn't until the F-12 that Farmall did away with the bull-gear drop box—except on high-crop models—and used larger tires instead to attain the desired clearance.

Due to their Culti-Vision design, Farmall A and B models used an extra-long clutch housing/torque tube.

A number of tractors, including this Farmall Cub, utilize an enclosed gear set at the wheel end of the axle to provide further gear reduction, while increasing ground clearance.

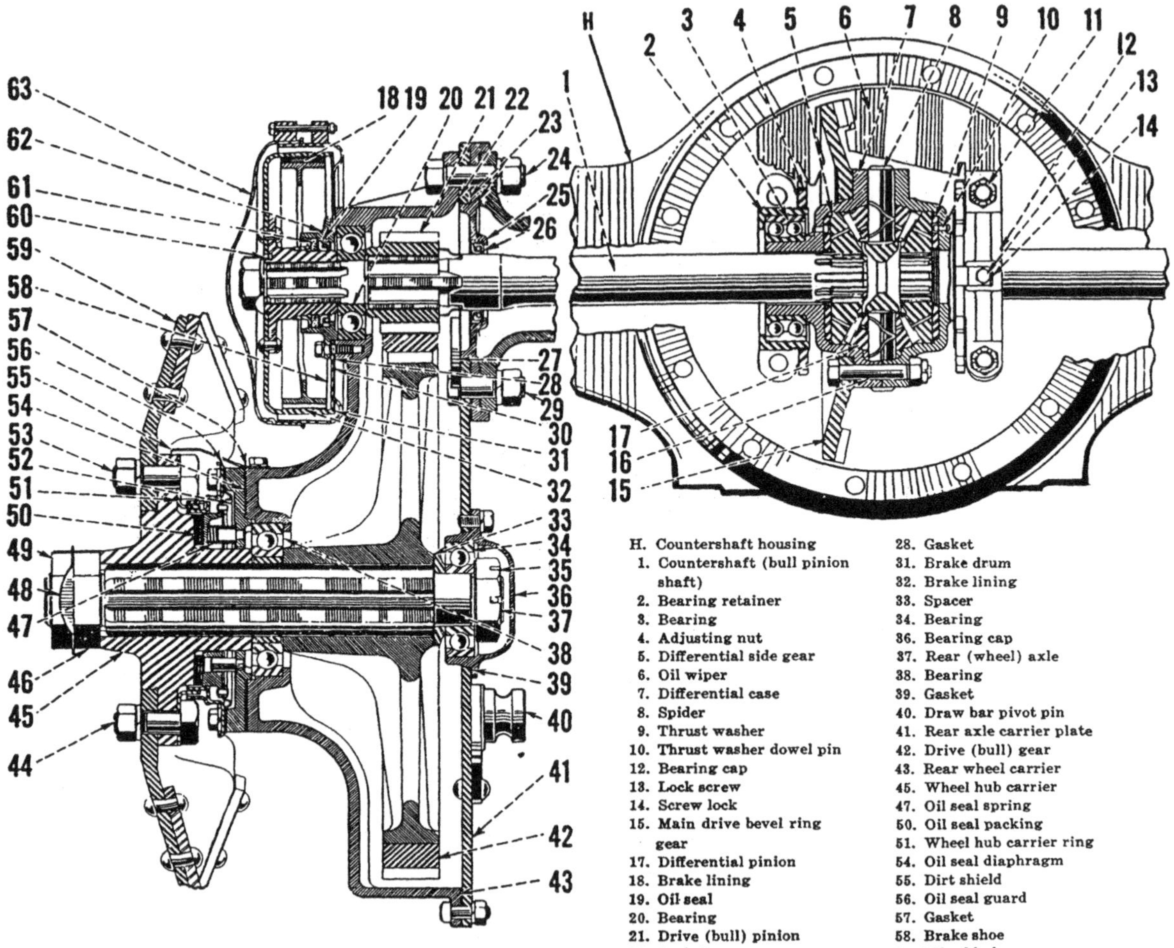

H. Countershaft housing
1. Countershaft (bull pinion shaft)
2. Bearing retainer
3. Bearing
4. Adjusting nut
5. Differential side gear
6. Oil wiper
7. Differential case
8. Spider
9. Thrust washer
10. Thrust washer dowel pin
12. Bearing cap
13. Lock screw
14. Screw lock
15. Main drive bevel ring gear
17. Differential pinion
18. Brake lining
19. Oil seal
20. Bearing
21. Drive (bull) pinion
22. Gasket
23. Gasket
25. End plate
26. Oil seal
27. Bolt lock
28. Gasket
31. Brake drum
32. Brake lining
33. Spacer
34. Bearing
36. Bearing cap
37. Rear (wheel) axle
38. Bearing
39. Gasket
40. Draw bar pivot pin
41. Rear axle carrier plate
42. Drive (bull) gear
43. Rear wheel carrier
45. Wheel hub carrier
47. Oil seal spring
50. Oil seal packing
51. Wheel hub carrier ring
54. Oil seal diaphragm
55. Dirt shield
56. Oil seal guard
57. Gasket
58. Brake shoe
59. Wheel hub
60. Nut lock
61. Felt washer
62. Bearing retainer
63. Mud shield

Fig. IH137—Model F30 differential main bevel drive ring gear, brakes, countershaft and rear axle assembly.

Above: This cross-section illustration of a Model F-30 shows the orientation of the spider gears, drive shaft, brake drum, and bull gear, which in this case is located within the drop axle housing. (Courtesy Jensales)

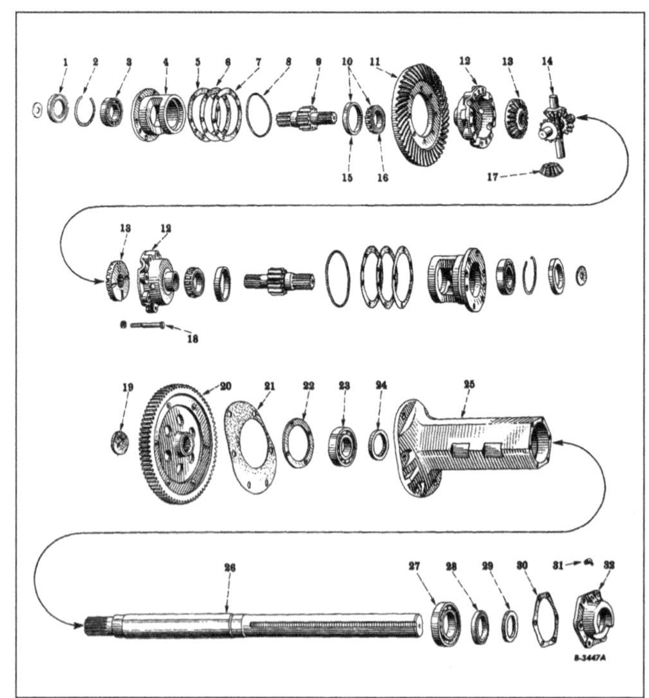

Right: This exploded drawing of a Model C rear housing and axle gives you an idea of what to expect when replacing the various seals and retainers.

Differential and Final Drive Inspection and Repair

Inspecting and rebuilding the differential and final drive is not too much different from working on the transmission. Basically, it means draining the old fluid, if it is different than that in the transmission, cleaning the gears, and checking for worn bearings and seals and missing gear teeth. Naturally, the different models and series built over the years used a number of different configurations, so you'll need to refer to your tractor service manual for adjustment procedures and end play or backlash tolerances.

On a number of tractor models, the gear mesh and backlash of the main drive bevel gears are controlled by shims on the shaft. In those cases, the repair manual generally recommends that tooth contact and backlash be checked and adjusted, if necessary, whenever the transmission is overhauled; it is imperative, however, when a new pinion or ring gear, or both, are installed.

If shim adjustment is in order, the first step is to arrange the shims to provide the desired backlash between the main drive bevel pinion and ring gear as specified in your tractor service manual. The next step is to adjust the shims to provide proper tooth contact or mesh pattern of the bevel gears.

To do this, paint the bevel pinion teeth with a substance called Prussian blue or with another called red lead. Then, rotate the ring gear in the normal direction of operation under no load and observe the contact pattern on the teeth surfaces. The areas of heaviest contact will be indicated by the absence of the coating. In other words, the paint will be removed from the points of high tooth contact.

Ideally, the contact area should be uniform from the top edge of the active profile to the lower edge where it breaks to an undercut. Don't expect to see much contact beyond the toe, or narrowest end of the tooth, though. The teeth are ground in such a way that they deform under heavy load, allowing the contact area to increase and move toward the heel, thereby increasing the load-carrying capacity of the gear.

In most cases, the final drive doesn't require anything more than a good cleaning and inspection of the gears.

If the heavy contact is concentrated high on the toe or near the outer edge of the tooth, the pinion generally needs to be moved toward the front of the tractor by adding a shim behind the pinion bearing cage. If, on the other hand, the heavy contact is concentrated low on the pinion tooth, you'll need to remove a shim from the pinion bearing cage. Refer to your manual for proper adjustment procedures.

Once you've obtained the ideal tooth contact, you'll need to recheck the backlash obtained in the first step to make sure it is still within the specifications.

One tip offered by restorers is that if you find bull gears with excessive wear, you can often swap them side for side. In effect, you're positioning the gears so that what was the gear lash in the forward direction is now the gear lash for reverse and vice versa. However, don't switch the bull gears without also switching the pinions.

A number of Farmall models, which utilize gear reduction at the end of the axle to increase torque, also employ a brake drum on the end of the axle shaft.

Axle Shafts

Like all other components in the final drive, the rear axles on most farm tractors were built tough. They still will need attention, though.

One of the most common problems restorers encounter is an oil leak where the axle exits the rear axle carrier. This is most commonly caused by a seal failure. With few exceptions, most models use at least two seals at the end of the axle housing; generally an oil seal and a felt seal. Of course, if the seal failed, there's a chance that the bearing needs replacing, as well. To check the bearings, raise the axle and, with the wheel off the axle and the transmission in neutral, check for excessive axle shaft end play by attempting to move the axle in and out and up and down.

Another potential problem is that the defective seal may have a groove worn into the axle. Not only will this contribute to the leak, but it will make it impossible to repair with a new seal. Hence, it's important to check the axle, just as you would any shaft on which a seal is being replaced. If you can feel the groove with your fingernail, it needs to be repaired.

While some shafts can be repaired with a Speedi-Sleeve, which is basically a thin collar that slips over the original shaft to create a new surface, a drive axle will generally require the second option. That is filling the groove with a metal-filled epoxy, such as J-B Weld. Two or three thin coats are usually better than one heavy one. Once it has cured, simply sand it to a smooth and symmetrical surface, finishing it off with extra-fine paper or emery cloth.

Even if the seals aren't bad, it's a good idea to replace them as part of the restoration process. Just remember to read and follow the instructions in your service manual. On some models, such as the W-9 and WD-9, the oil seal is installed with the lip facing toward the bearing. And on others, like the rice versions of those same tractors, the seal is installed with the lip facing away from the bearing.

Brake Restoration

Let's be honest: The brake system is not the place to cut corners. You might be able to get by without overhauling the transmission or opening up the rear end, but you'll want to give the brakes the attention they deserve. This isn't just for your safety, but for the safety of those around you. If you take your tractor to any shows at all, you're going to need to unload it off a truck or trailer. And this is not the place to have your brakes fail to hold. Consider, too, how you would feel if you lost control of your tractor in a parade, with people lining both sides of the street. Fortunately, brake restoration is not a difficult job.

Although you'll still find a number of vintage farm tractors with only one brake, which was primarily used for parking or holding the tractor in place while it was being used as a belt pulley power source, Farmall tractors were among the first to have a brake for each rear wheel. Farmall was also unique in its use of an automatic braking system that applied the brake on the rear wheel in the direction in which you were turning. This was accomplished via a cable that tied each brake to a bracket on the steering pedestal. As the wheel was turned, it pulled on the cable and automatically applied the brake.

For the most part, though, Farmall and McCormick-Deering tractors used one of two types of brakes. The first type used externally contracting band brakes that contracted on drums mounted on the differential or bull pinion shafts.

The other type used a double disc, self-energizing system that was splined to the outer ends of the bull pinion and integral brake shaft. The brake disc assembly includes two separate discs and a pair of actuating discs that are connected by three springs and held apart by five steel balls in matching grooves. When the brake pedal is depressed, the linkage cause the two actuating discs to rotate slightly in opposite directions. This causes the balls between the actuator discs to roll to the shallow end of the grooves. This, in turn, forces the brake lining discs to come in contact with the stationary braking surfaces. Since the brake discs are splined to the bull pinion, the rotating force on one of the actuating discs—depending on which direction you're traveling—helps force the balls up the ramp, making the brakes somewhat self-energizing—in effecting providing some of the benefits of power brakes before power brakes were ever available on a farm tractor.

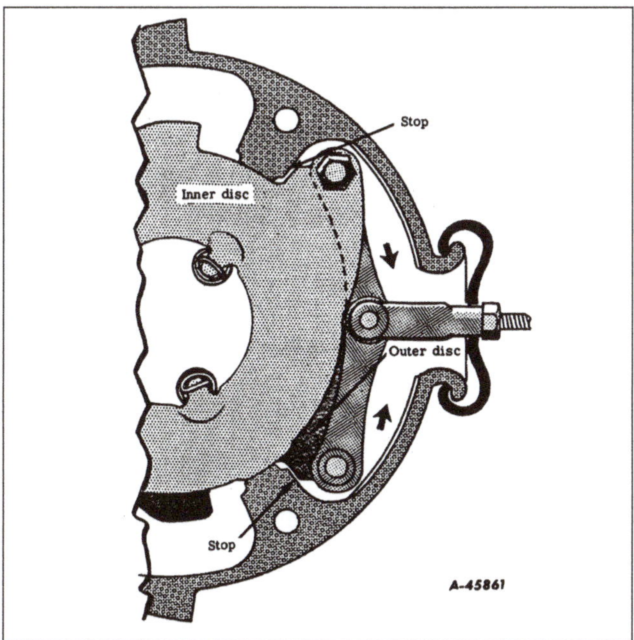

This side-view drawing of IH's double disc brake system illustrates the operation of the actuating links and how they rotate the actuator discs in opposite directions.

Due to rust and wear, one of the brakes on this Farmall Regular required a new drum. Note that because of the drop axle, the brake drum is on the end of the drive shaft instead of the bull gears.

Francis Gordon, from Agra, Kansas, disconnects the brake pedal for brake restoration on a Farmall H.

Brake Repair

Before you begin any brake repair project, remember that brake springs can fly off unexpectedly and in who-knows-what direction if they are not handled properly. Always wear safety glasses when attempting any work on the brakes. Since you will no doubt be lifting the rear of the tractor off the ground, you should also double-check to make sure it is secure from rolling or tipping.

While it's important that you start brake troubleshooting by checking all brake adjustment points, it's also important to examine the brake linings or discs, as well. If brake linings or discs are glazed but aren't badly worn or oil soaked, you may be able to bring them back to life by simply roughing them up with sandpaper. Other restorers have used a torch to dry out oil-soaked brake shoes that have sufficient wear left on them. Keep in mind that brake pedal pressure required to obtain sufficient braking action will increase as the lining or discs in particular wear thinner.

If you do find that the brake linings or discs have become oil soaked or worn too far, it should be fairly easy to find replacement parts. If a brake kit is no longer available, any automotive shop that does brake work should be able to rivet new linings in place on your brake shoes. The same shop should also be able to turn any problem brake drums on a lathe, removing any grooves or out-of-round spots. In the meantime, it's worth noting that the left and right disc brake assemblies are the same, which means that all parts are interchangeable in the event you need to find replacement parts.

Once you've checked the pads or discs, you need to make sure the brake drum or disc brake contact surfaces are clean and free of rust. But again, a piece of emery cloth or sandpaper will fill the bill. On disc brakes, it's important to also check the actuator disc ramp for wear or rust that could prevent the ball from rolling up the ramp. Finally, check to make sure the actuating disc balls are not rusty or out of round. If they can't be polished to an effective state, they should be replaced.

The next step is removing the brake housing cover that incorporates the band assembly.

The band brakes used prior to the advent of the double disc system contracted on drums mounted on the outer ends of the bull pinion shafts.

If you're lucky, you may be able to bring the brake band linings back to life by simply roughing up the surface with sandpaper, or using a torch to carefully dry out oil-soaked pads. Otherwise, you'll need to replace the brake band lining and perform a thorough clean-up.

Splines on the end of the bull pinion and integral brake shaft provide the braking points on International Harvester's unique double disc brake system.

Because the inside of the brake housing serves as one of the braking surfaces, it's important that it be cleaned as part of disc brake restoration.

Notice how the braking surfaces on both the brake housing and actuating discs have been cleaned of rust.

New disc brake discs are ready for installation on the brake shaft.

Roy Troxel, who works as a mechanic for Farmall enthusiast Bob Devling of Elwood, Kansas, reassembles the actuation discs.

As a final step, Troxel coats the balls and ball ramps with a lubricant/rust preventative prior to assembly.

On some later-model Farmall tractors, it's possible to remove the bull pinion/brake shaft as part of brake overhaul without opening the final drive cover. In this case, the pinion was replaced because wear on the splines were causing the brakes to bind.

Brake Adjustment

Whether the brake's pads or discs have been renewed, touched up, or approved in the current condition, it is important that the brakes be adjusted. On Farmall tractors, which generally use foot brakes that are adjacent to each other so they can be locked together, it's especially important that brakes are adjusted to a comparable setting with an equal amount of free travel. Otherwise, one brake may be applied while the other just drags. Unequal braking action is generally indicated by the tractor pulling to one side when the brakes are applied simultaneously. This isn't a problem, of course, with hand brakes or on tractors where the left and right brake are operated by the corresponding feet.

The easiest way to adjust the brakes in most all cases is to raise the rear wheel off the ground and tighten the brake until you can no longer rotate the wheel by hand; then back it off until the rear wheel turns with only a slight drag. Naturally, the adjustment process will vary depending upon the tractor model and brake configuration. Quite often the adjustment instructions will also tell you to depress the brake pedal approximately 2 inches and adjust the brake until the lining contacts the drum as a starting point.

At the least, you should end up with around 1 to 1 3/8 inches of pedal free play before the brakes make contact.

After the brakes have been refurbished and reinstalled, adjust the brake to the point the rear wheel begins to drag and back it off slightly.

Foot brakes that are adjacent to each other, or can be locked together, need to be adjusted to a comparable setting with an equal amount of free travel. This will prevent the tractor from pulling to one side when both brakes are applied in an emergency.

Chapter 9

Front Axle and Steering

As was the case with nearly all tractor models, the first International Harvester and McCormick-Deering models were equipped with wide front axles only. This was basically a carryover from the steam-tractor era in which tractors were used for plowing and belt applications. Even at the time the McCormick-Deering 15-30 Gear Drive tractor was introduced in 1921, tractors were being used for wheatland-type applications, like pulling a plow, disc, or grain binder, and for rugged jobs like pulling stumps. Cultivation was still done by a team of horses. Of course, belt applications were common, which fit the IH 10-20 and 15-30 machines just fine.

However, in 1924, International Harvester changed more than tractor design when it introduced an "all-purpose," or tricycle-type, front end on the McCormick-Deering Farmall. The new design changed the entire tractor industry. By the 1930s, nearly every manufacturer had introduced a tricycle configuration as standard equipment on certain models and as an option on others. To accommodate the needs of tractor owners, most companies, including International Harvester, even offered a choice of single- or dual-wheel tricycle front ends. By the time World War II ended and many veterans were returning to the farm, some dealers had a hard time selling a tractor with a wide front end.

Not only did the tricycle front end fit between corn and vegetable rows of the time and make shorter turns at the ends of the field, but most farmers also liked the visibility the narrow front axle offered when using a front-mounted cultivator. In fact, the Farmall was one of the first to use its tricycle design as a base for other implements, like a two-row corn picker that mounted directly on the tractor.

Unfortunately, tractors with a narrow front axle also proved to be more dangerous than those with wide axles, especially if they were top heavy or turned too fast on sloping terrain. Times were changing, too. As equipment got bigger and the turning radius improved on standard tractors, farmers found they didn't have to turn such tight corners. They also discovered that they could guide the tires on a wide front axle between the rows as easily as they could keep dual tricycle wheels in a single row width. Besides, rows were getting narrower. Consequently, tractors today have gone back to wide front axles exclusively.

Steering Configurations

The steering systems on early Farmall tractors were pretty simple in comparison to today's power-steering systems. The ultimate picture of simplicity, the steering shaft on the first Farmall Regular had a steering wheel on one end and spines on the other end, which connected to an open gear head at the top of the steering pedestal.

Through most of the Farmall and McCormick-Deering production era, though, International Harvester tractors were equipped with one of two types of steering mechanisms. The tractor either utilized a pedestal with a steering sector at the top of an enclosed steering spindle; this is found on virtually all F Series models, as well as most of the larger row-crop tractors, including the M, H, 300, 400, 350, and 450. Or the tractor employed a steering gearbox that connected to a steering shaft, which, in turn, connected to a steering arm; this is found on all McCormick-Deering wheatland-type tractors, such as the W-6, WD-9, etc. The A, B, C, and Cub tractors also utilized a steering shaft that angled down to a gearbox, rather than the pedestal configuration.

Regardless of the configuration, restoration will primarily consist of replacing worn parts, bearings, bushings, and seals, and/or making the appropriate adjustments. This will include adjusting for wormshaft end play; vertical shaft or vertical spindle end play, depending upon the model; and backlash. On some models, the camshaft end play and gear backlash are adjustable and on others, it is not, since the worm shaft bushings are pre-sized. On the M and H, for example, end play of the upper bolster pivot shaft is nonadjustable and vertical thrust is taken on the ball thrust bearing. Your service manual will provide the proper technique for overhaul, as well as the specifications for adjustment. In each case, though, it will be necessary to support the front of the tractor to remove the load from the steering gear.

The steering gears on the Farmall Regular were pretty basic. In fact, they weren't even enclosed in a housing. The open gears simply connected the steering shaft and steering pedestal.

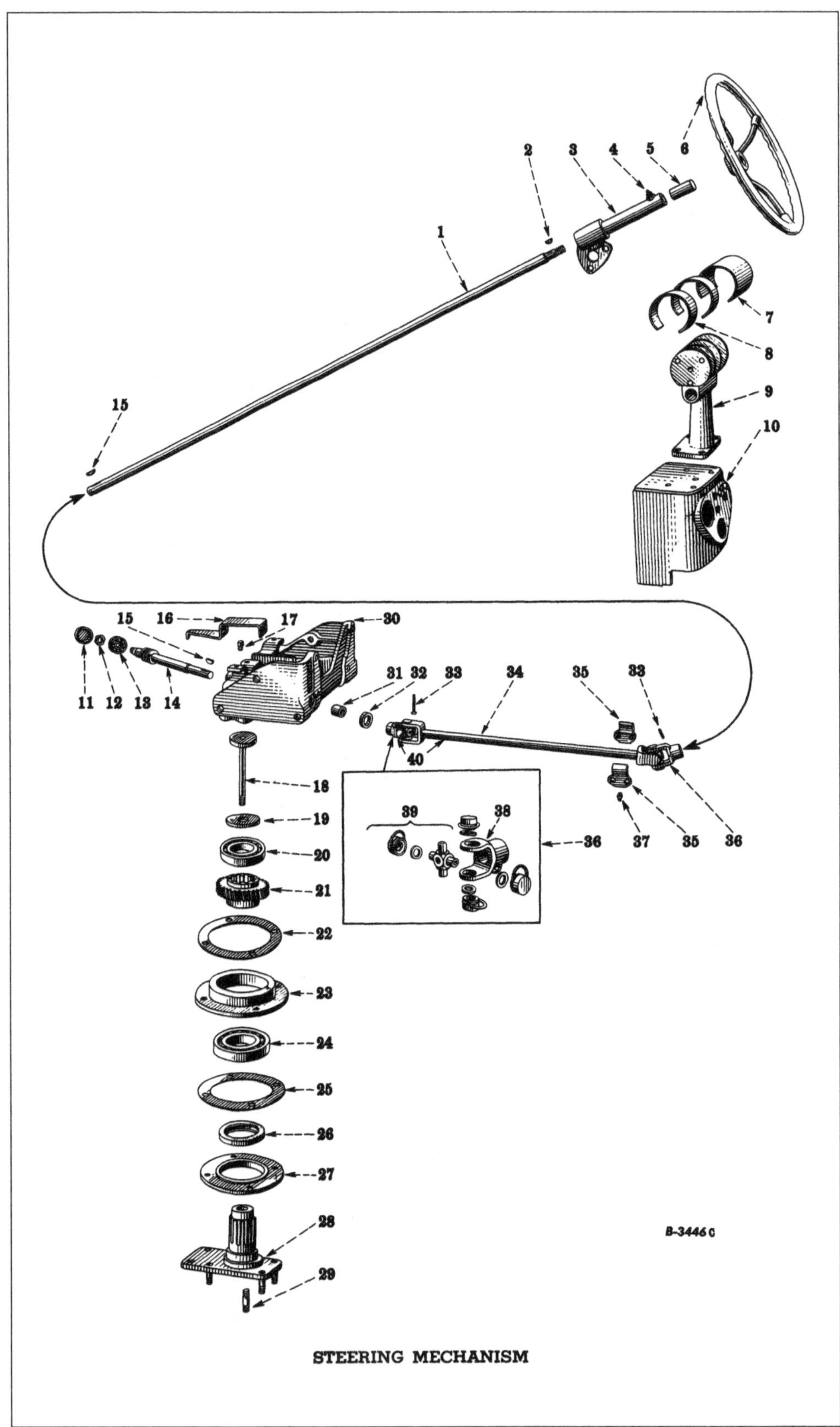

STEERING MECHANISM

While it differs from the typical pedestal steering assembly, the steering mechanism on the A, B, C, and Cub operates in a similar manner via a worm and gear drive.

One of the next steps in the Farmall steering system evolution was a housing often referred to as the Duckbill. This system was used on some early F-20 and F-30 models prior to the development of the worm and gear arrangement.

Eventually, Farmall moved to a more conventional gear set housing on the majority of F Series models.

Above: Farmall was unique, not only in having the first row-crop tractor, but in its development of the steering wheel/cultivator gang shift which steered a cultivator through the field as the front wheels were turned.

Right: A latch on the implement, which engaged a tang on the steering post, could be disengaged from the operator's platform.

Several models, including utility versions and the Cub, A, B, and C used a steering gear housing bolted to the front face of the engine, rather than the bolster system used on other row-crop models.

By the end of 1929, the tricycle-type tractor accounted for 40,000 units out of 229,000 produced by 47 tractor manufacturers.

Although several tractor companies, including IH, produced tricycle models with a single front wheel, they still tend to be rarer than models with dual tricycle wheels. This one was obviously set up for towing.

One of the steps in steering system restoration is checking for free play in the bolster on narrow front models. The excess grease around the top of this bolster also indicates the likelihood of a faulty seal.

Front Axle Repair

The life a vintage tractor led before you acquired it has a lot to do with its condition and the repairs that it is going to need. The condition of the steering gear and front axle is a textbook example. A tractor that spent most of its days in large wheat fields in western Kansas, for example, isn't going to have near the wear on the knuckle bushings and axle pivot pin as a row- crop model that spent every working hour crossing corn furrows and turning around on end rows. This is just one more reason to find out all you can about the tractor before you make a purchase or calculate repair costs.

That said, the only satisfactory way to overhaul the front-axle assembly on a wide-front tractor is to remove it from the tractor and make a complete check of all bearings and bushings. That includes the center pivot pin, steering arms, ball seats on tie rods, knuckles, and wheels. It's important to note that on most tractors, the knuckle post bushings are pre-sized to provide a specified amount of clearance for the knuckle post.

On models with a tricycle-type front end, inspection primarily consists of checking and replacing wheel bearings and seals. In fact, on most models, it's possible to remove the axle assembly or bolster assembly from the bolster, making repair or replacement of the axle even easier. Depending upon the amount of free play in the bolster, it may also be necessary to replace bushings, bearings, and seals in the upper or lower sections of the bolster.

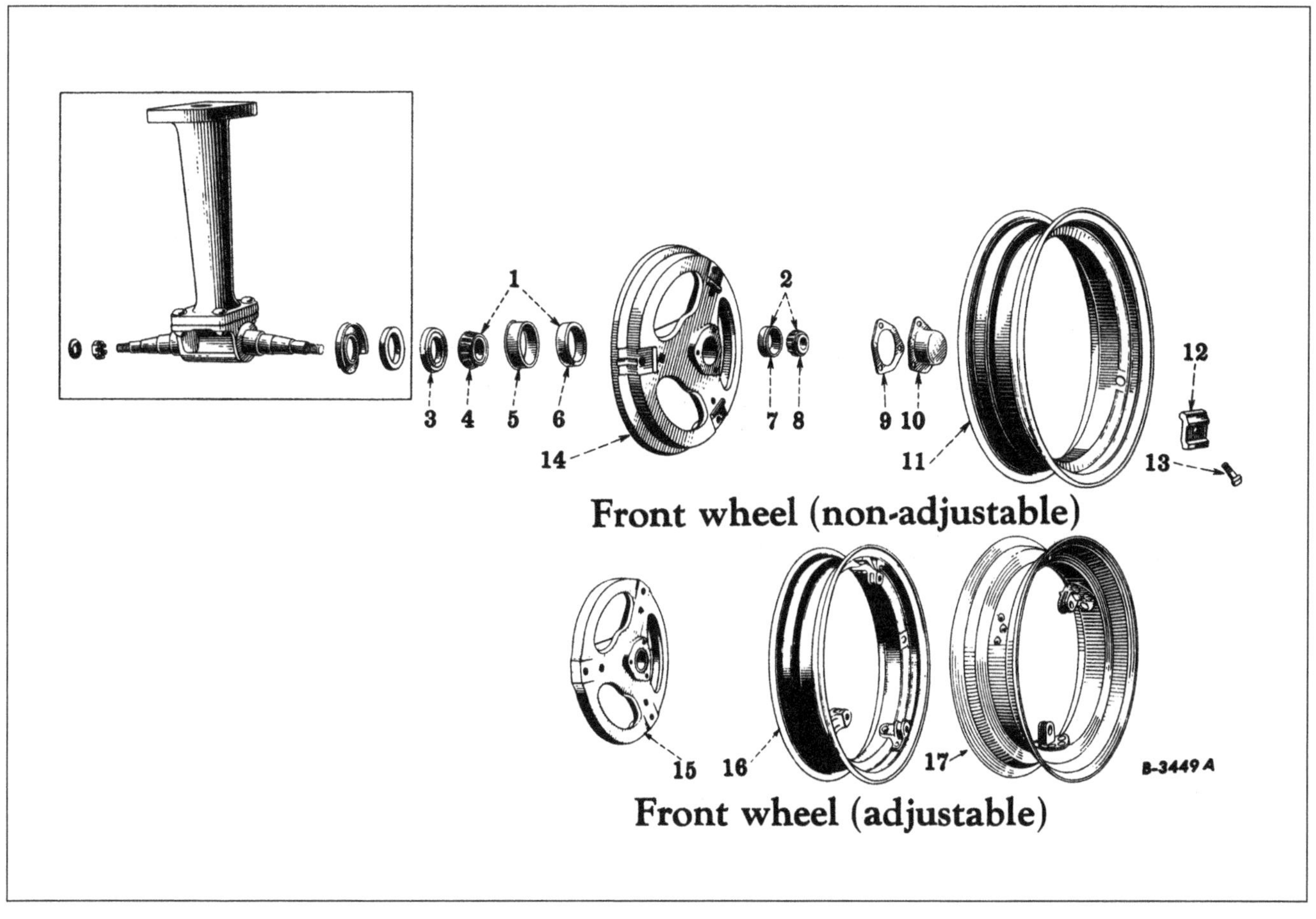

This exploded view of the front axle and bolster on a Farmall C illustrates the various bearings and seals that should be checked during front axle restoration.

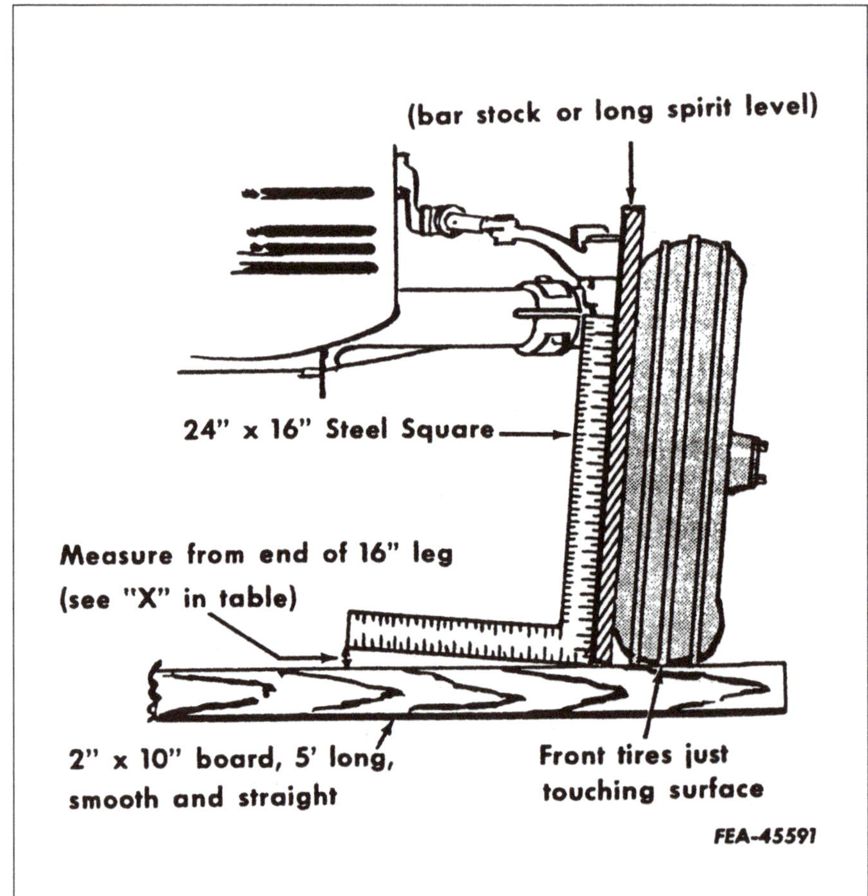

Your service or repair manual should provide instructions on how to check and adjust the front wheel camber on wide-front axles. (Courtesy of Jensales).

Thanks to the number of mounting holes on the frame, it's pretty easy to make a set of stands to support the front end for axle repairs.

Since the lower portion of the bolster can be removed from the shaft, it's not necessary to remove the entire assembly from the housing unless other repairs are needed.

Walter Bieri designed his own tool for removing the bolster shaft from row-crop models. He simply welded a shaft to a nut that fits the threads on the top of the shaft. Then he can use a large hammer to drive the shaft out of the housing without risk of damaging the threads.

The steering gear unit can be disassembled and overhauled without removing the steering gear housing from the tractor.

With few exceptions, most Farmall models use at least two seals at the end of the axle housing.

Walter Bieri, a Farmall collector from Savannah, Missouri, removes the lower seal and bearing from the steering housing on a Super M.

With the steering assembly overhauled and reassembled, Bieri is ready to reinstall the wheels and remove the support stands.

Power Steering

As with the hydraulic system, the maintenance and repair of the power steering system requires absolute cleanliness of all parts. It's also important that all parts be free of nicks, burrs, or scoring that can affect performance. A control valve and motor is the heart of the hydraulic steering system. In fact, the unit is mounted on the steering shaft itself with the shaft passing through the center of it. The control valve's role is to direct the flow of fluid to the proper side of a hydraulic motor that essentially assists the manually operated system. Hydraulic fluid is supplied to the cylinder from an engine-driven pump and is returned to the tractor hydraulic system reservoir from the motor.

To inspect and service the system, it's best to follow the procedure outlined in the tractor service manual, since service involves checking pressures; adjusting flows; and inspecting various valves, washers, springs, and end play clearances.

Above: The power steering unit on Farmall tractors is basically a fluid-operated servo-mechanism that employs a gear-type motor mounted on the steering shaft.

Right: Over the years, other companies offered attachments and add-ons to improve the performance of customers' tractors. One of those was Char-Lynn, which offered a field-installed power steering unit.

Steering Wheel Repair

Unless your tractor has been protected from the elements for most of its life, there's a pretty good chance the steering wheel is going to be severely cracked. Fortunately, there are several solutions available to you. Should you prefer to have the steering wheel professionally repaired, there are several companies that will take your old wheel and mold new plastic around the steel rim, complete with the original grooves, ribs, or finger ridges.

Should you choose to repair a cracked steering wheel yourself, there are a couple of options practiced by restorers. One professional restorer says he uses Fiberstrand body filler to fill all the cracks and crevices. Another uses a body filler such as Evercoat polyester glazing material and follows that with a coat of fast-fill primer. With either product, the steering wheel must be sanded smooth after the material hardens and then painted.

You can always buy a reproduction or refurbished steering wheel as well. Just be sure you get the right style for your particular model, assuming authenticity is important to you.

It won't take much more than a sheet of sandpaper and some paint to refurbish the cast steering wheel on this older IH model.

For a professional look on a hard-rubber-rimmed steering wheel, you might consider sending it to one of several companies that specialize in refurbishing plastic or wood rimmed steering wheels.

Chapter 10

Tires, Rims, and Wheels

Tires and wheels can be a costly and frustrating part of a restoration project. If your vintage Farmall tractor was originally equipped with steel wheels, there's a good chance that rust has taken a toll. However, there's a more likely chance that a previous owner or backyard mechanic removed the steel rims so the tractor could be fitted with rubber tires. Although some shade-tree mechanics went to the trouble of cutting the steel rims off the spoked wheels before welding the spokes to drop rims designed for rubber tires, others simply ordered new wheels through their International Harvester dealer.

Rubber-tired wheels were first offered on the majority of Farmall models in 1936, although one source claims that IH offered rubber tires for the first time in 1934 on the F-20. On the other hand, customers could still order a tractor with steel wheels all the way into the 1950s. As was the case with all tractor brands, steel wheels were particularly prominent during the World War II years—that is, assuming you could even buy a tractor. Hence, you may need a little help finding out what was original, what was optional, and what was engineered in the field.

In the meantime, you'll need to decide whether you want to keep the rubber tires or go back to steel wheels to attain the original look of the tractor.

That's not the last of your decisions, though. If the tractor was equipped with rubber tires from the factory, chances are the tires have been replaced with different tires than originally came on that model. If you're planning to use the restored tractor as a work tractor, that may not matter to you. But if you plan to restore it as a show tractor, you may want to find the correct components.

Rear Wheel Removal

Most row-crop Farmall tractors are equipped with an adjustable rear axle that allows the wheels to be moved in or out for variable tread widths. If you're lucky and the tractor has been well cared for, you should be able to slide the rear wheels on the axles, or even off the axles, in the intended manner.

Unfortunately, wheel hubs are like a lot of components on classic tractors. If they have sat long enough without being moved, there's a good chance they have solidly rusted into place. If this is the case, you'll need to start the removal process by saturating them on a regular basis with some type of penetrating oil.

If after trying to remove the hub in the traditional manner, it still doesn't budge, you may need to heat the hub with a torch to expand it. Don't expect a propane torch to do it, though. The hub will absorb a lot of heat, so an acetylene torch is usually more appropriate. Don't beat on the hub when it's hot. This will only deform the parts and make matters worse.

Once the wheel and hub have been removed, by whatever means necessary, you can clean and smooth the axle so that the wheel will slide effortlessly when reinstalled.

Considering that whitewalled snow tires were never used as stock equipment, this tractor will at least need new front tires.

Ever wonder why some steel wheels are so hard to find? In an effort to convert steel-wheeled tractors to rubber tires, shade-tree mechanics often welded tire rims to the old wheel or spokes, as was the case with the identical wheels on the right.

Tire Repair and Restoration

If the tractor has been sitting for some time, it's quite possible that the rims have been corroded by calcium chloride. Or perhaps the tires have rotted away after years of sitting in a pasture and the wheels came into contact with the ground, allowing rust to take its toll. If you need one more thing to worry about, it's the fact that replacement tires can be rather expensive, especially if you're restoring a behemoth like the International 600. A new set of its 14.00x34 rear tires can easily set you back at least $700.

If you're only looking for a working tractor, any tires that fit the rim will generally be acceptable. The newer style tires, with their 23-degree bar, or long-bar/short-bar design, may even offer better traction than the 45-degree lug tires originally found on the tractor.

However, if your goal is to restore a vintage model to show condition, and you're after accuracy, the challenge is a little greater. Not only did the tire companies change their size standards, but they also changed tire styles as more effective patterns were discovered. Fortunately, there are several independent sources for the most sought-after sizes and types of tractor tires. They include M. E. Miller, Gempler's, and Wallace W. Wade.

One option for repairing tires that are worn or slightly damaged, but still usable, is to install a set of tire reliners. These are made from old tires that have had the lugs ground down, making the reliner itself about 1/4-inch thick. Generally, reliners come in half-moon shapes so they can be easily inserted into the old tire, where they partially overlap. Some companies also offer spot reliners. The inner tube is inserted into the tire, where, once inflated, it holds the reliners in place.

Other tire repair products include rubber putty that can be used to repair cracks, gouges, weather checking, and other minor problems. Unfortunately, it can't be used to repair a hole.

Finally, various companies carry a concentrated black tire paint that can be used to revive the color of old, gray-looking tires. Simply mix it with paint thinner according to the directions on the label and apply it as you would paint.

Although tire reliners can be used to salvage some tires, the hole in the sidewall on this rear tire has essentially ended its usable life.

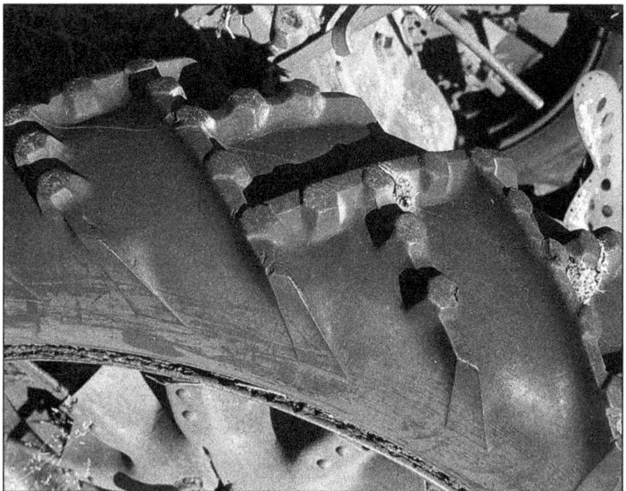

It's next to impossible to find tires with this type of vintage tread. Too bad this one is too far gone to reuse.

If you're restoring a work tractor, newer style tires, with their 23-degree bar generally offer better traction than the 45-degree lug tires originally found on the tractor.

The addition of an outer rim that extends beyond the lugs is one way to make a steel-wheeled tractor parade ready.

Wheel and Rim Restoration

As you'll quickly discover, International Harvester used a variety of wheel sizes and types on their tractors over the years. In addition, a variety of lug types were offered in an effort to match the application. Some aftermarket equipment, such as the F&H round-spoke wheels, were even offered as an option. As a result, there can be a lot of confusion about what wheels should be considered original.

Front wheels, meanwhile, went through a similar evolution. Early options for many row-crop tractors included cast disk wheels, as well as spoked steel wheels that could be fitted with both guide bands and extensions. The former not only helped the front end track through turns, but, with the aid of even higher guide rings, also helped the operator hold the tractor on the top of listed corn rows.

If you've been fortunate enough to find rims and wheels that are in relatively good shape, the restoration process may be as simple as sandblasting the appropriate parts and applying one or more coats of primer prior to painting.

Unfortunately, vintage wheels and rims don't take kindly to age, especially if they were equipped with rubber tires and ballasted with calcium chloride. They don't take well to rust, either. If one or more of the wheels are too far gone, or if a rim is in very bad shape, the best bet—particularly if you're dealing with a clamp-on rim—will be to frequent the auctions, swap meets, salvage yards, and classified ads for a replacement. Farmall collectors are fortunate in that International Harvester started using clamp-on rims earlier than most manufacturers and had them available almost as soon as rubber tires were offered.

If you're only dealing with a few rust holes on a rim, you should be able to take care of those by thoroughly cleaning the holes and filling them in with small beads of weld or some type of filler, such as J-B Weld. You'll want to use a filler primer before painting to further smooth imperfections.

If you're good with a welder, you may be able to repair larger rust areas by totally replacing the damaged area. The first step will be cutting out the corroded or rusted portion of the wheel. Quite often, this will be the outer edge of the rim, where it has rested on the ground. Next, you'll have to find a scrap wheel or rim that is identical in size and style, from which you can cut a replacement piece. Make sure all edges have been ground smooth, clamp the splice into position, tack weld around the whole piece to keep it from warping, and carefully weld it into place. Once you've finished welding, grind all splices down until they are flush with the surrounding metal and prepare the wheel for painting.

Another option is to have new rims installed on your wheels. Detwiler Tractor Parts in Spencer, Wisconsin, offers this service on several spoked and pressed steel wheels. Nielsen Spoke Wheel Repair in Estherville, Iowa, is another vendor that will put new rims on the centers of your old wheels.

International Harvester used a number of different wheel styles—with and without spokes—during the years steel wheels were standard or optional.

This rear wheel is beyond restoration, but pieces of the rim could still serve as a repair or splice on a similar rim. Due to the value of the spoked wheels, a restorer might also look at getting a new rim installed on the spokes.

Skeleton wheels were just one of the steel wheel options Farmall offered on row-crop models. Many collectors find the thin wheels particularly appealing.

International Harvester used a variety of wheel sizes and types on their Farmall tractors over the years, which can lead to a lot of confusion about what wheels should be considered original.

Plenty of Farmall collectors would love to get their hands on this stack of steel wheels nearly overgrown by brush.

Neglect and years of exposure to calcium chloride as rear wheel weight have obviously taken a toll on this wheel.

Somebody, years ago, did a good job of welding a set of rims onto the spokes from a set of steel wheels. These were found on an old 10-30 that was built long before rubber tires were an option.

Chapter 11

Hydraulic System

Depending upon the age of your tractor, you may not even need to worry about a hydraulic system. After all, hydraulics weren't found on many early tractors. In fact, it wasn't until the late 1930s, when the Ford 9N combined the Ferguson hydraulic-lift system with the three-point hitch that hydraulic force was used on tractors to raise and lower implements and attachments.

International Harvester, however, took a couple of different directions on its own. Due to the popularity of front-mounted and mid-mounted implements on Farmall row-crop tractors, IH innovations began with its unique Touch-Control system, introduced on the Super A and Model C. This system allowed the operator to individually adjust the left and right sides of the mid-mounted lift. In addition, the system used live hydraulics, meaning it was constantly powered off the governor/ignition drive.

The other International Harvester innovation was the Fast-Hitch system, which was first introduced on the Super C in 1953. Marketed as the "click and go" hitch, it offered important implement draft control features, as well as quick mounting and dismounting. Fast Hitch implements were essentially equipped with two prongs that inserted into two receptacles on the tractor hitch. As they slid into position, they snapped into locks that held the implement in place. The operator could even unhook implements without ever leaving the tractor seat. Sales literature and magazine ads insisted that all the farmer had to do was back in and go. Unfortunately, as convenient as the system was, IH couldn't compete with the widely used three-point hitch system that was gaining ground. Part of the reason might have been that only IH produced Fast Hitch implements, which limited availability of Fast-Hitch equipment. At any rate, early Fast-Hitch tractors, and implements to go with them, are frequently sought after by Farmall collectors.

Basic Principles

Before we look at troubleshooting and repairing the hydraulic system, let's look at some of the basics associated with hydraulic systems. First of all, hydraulic fluid is just like any other liquid—it has no shape of its own and acquires the shape of the container. Because of this, oil in the hydraulic system will flow in any direction and into any pump or cylinder, regardless of the size or shape.

Like any fluid, it is also practically incompressible. As a result, when force is applied to hydraulic fluid, it transfers force to the work site. Hydraulic fluid has one other characteristic, though. It has the ability to provide substantial increases in work force, meaning that one pound of pressure on the piston in a small pump is converted to several pounds on a larger cylinder or piston. If you use a hydraulic bottle jack to lift your tractor, you already know how this works.

Now, let's look at the application of this principle on your tractor. Instead of pressure being supplied by a jack handle and piston, it comes from a hydraulic pump. From there, it flows through a valve, which directs its path to the appropriate point, and finally to a hydraulic cylinder on the three-point hitch, power steering, or remote.

Most hydraulic pumps used on farm tractors today are one of three types: a gear, vane, or piston pump. However, you're most likely to run into a gear pump, since it is the most common type of pump used in both hydraulic and power-steering systems on early Farmall tractors.

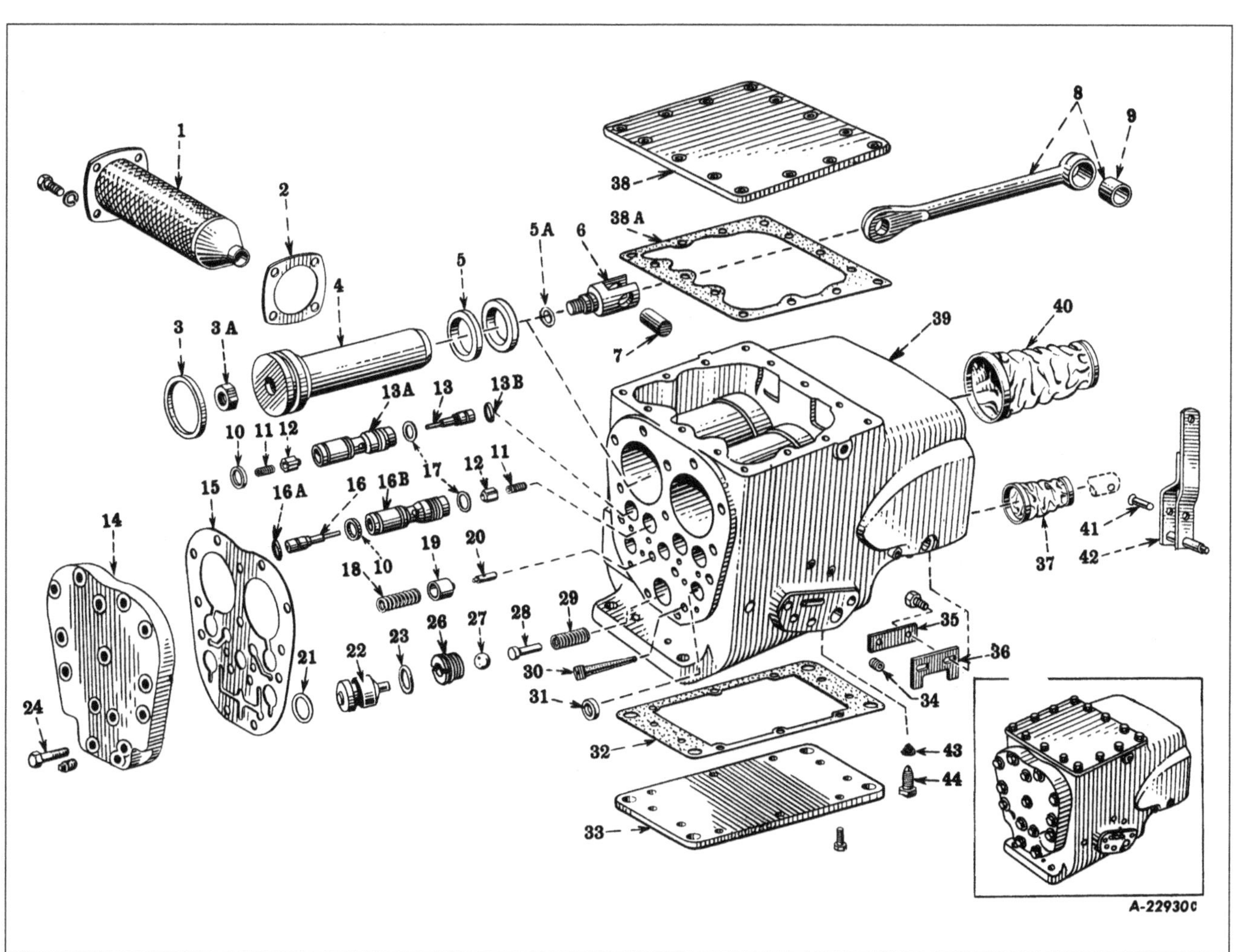

As illustrated in this exploded view drawing, there are a number of seals, bushings and springs in the Touch-Control cylinder block that are subject to wear.

Hydraulic System Contamination

The two biggest enemies of a hydraulic system are dirt and water. If these two contaminants were kept out of the system by the previous owner, you may not have any problems with the hydraulic system. But if they managed to work their way into the system, you may have some repairs ahead of you.

Just as in the engine, dirt can score the insides of cylinders, spool valves, and pumps. Water, meanwhile, will break down the inhibitors in the hydraulic oil, causing it to emulsify and lose its lubricating ability—again, leading to scoring of cylinder walls and breakdown of internal seals. Unfortunately, the tolerances in many hydraulic pumps and spools are even tighter than those in an engine.

One of the main ways dirt enters a system is through the air breather in the reservoir. The air breather is designed to let air move in and out of the reservoir in response to changes in the fluid level. It is also supposed to screen out dust by trapping it between layers of oil-saturated filter material. Unfortunately, many farmers failed to clean the filter or check for cracks or leaks that permitted dirty air to penetrate the system.

Keep in mind, however, that many older tractors do not have a special reservoir or cooler for the hydraulic system, but simply use the same oil that lubricates the transmission and differential.

Another way dirt can get into the system is through the careless handling of the hoses, particularly if a

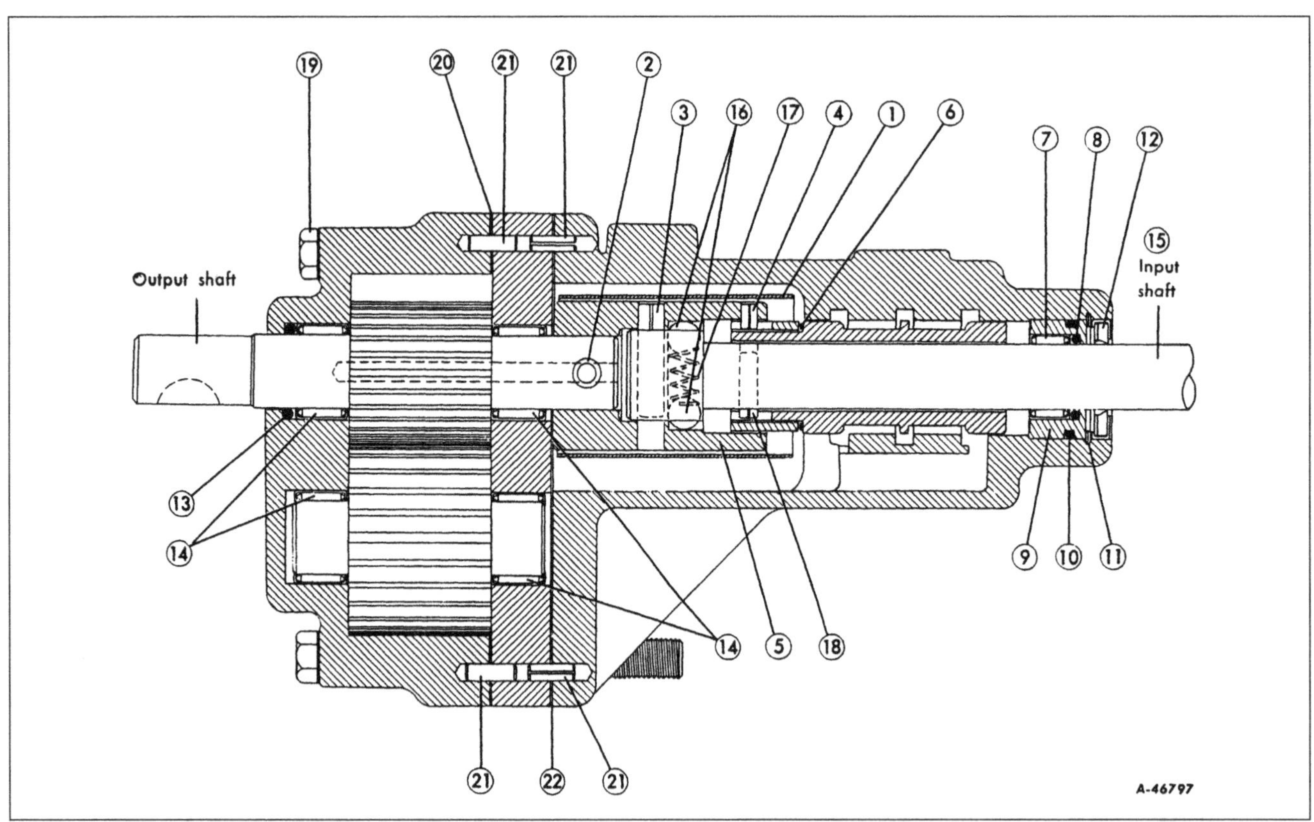

This illustration of the power steering unit on Farmall row-crop tractors helps explain its operation. Basically, turning the steering wheel activates the control valve which directs fluid to the appropriate side of the hydraulic motor. The motor, in turn, simply boosted or assisted the operator's effort. (Courtesy of Jensales)

broken or damaged hose is replaced. This should be a hint to practice cleanliness when making hydraulic repairs or inspections.

Dirt isn't the only enemy, though. Sludge, which is formed by the chemical reaction of hydraulic fluid to excessive temperature changes or condensation, can also cause havoc. If enough sludge builds up on the pump's internal parts, it will eventually plug the pump. To add insult to injury, a restriction on the inlet side of the pump can starve it of fluid, and heat and friction will cause the pump parts to seize.

Troubleshooting

Like many other systems on the tractor, you should have an idea whether the hydraulic system actually works and how well it works from your initial test drive, assuming that the tractor was in running condition when you acquired it. If the tractor has power steering, did it work? Did the lift system raise and lower properly?

Depending upon the age of your tractor, you're probably lucky to have either one, although International Harvester offered a hydraulic lift on its tractors as early as production of the F-12. If anything, you may just have a single remote outlet for attaching a set of hydraulic hoses.

If your tests indicate a weakness in the system, the first thing you should do is check the fluid level in the reservoir and make sure it is filled to the proper level with the recommended grade and type of fluid. Improper fluid can not only cause low or erratic pressure, but it can eventually deteriorate seals and packing, particularly if it contains incompatible ingredients.

Next, check for any problems with the hoses, including kinking or leaks. Keep in mind that hydraulic pressure escaping under high pressure through a pinhole leak can actually penetrate the skin, leading to gangrene poisoning if not treated quickly. Hence, you should never check for leaks with your bare hands, or even with leather gloves. Instead, use a piece of cardboard or wood passed over any suspected areas to check for escaping fluid.

If you have access to a pressure gauge or know someone who can assist, you can also check the pressure in the system to see if the problem is in the pump or in one of the valves. Most repair manual procedures call for inserting a 1/2-inch pipe tee in series with the pump discharge hose or the outlet elbow at the pump. A pressure gauge, capable of registering at least 1,500 psi, is then installed in the tee. A service manual for the Farmall A, C, and Cub instructs troubleshooters to even attach wheel weights to the rear rockshaft arms before taking a pressure reading. Whatever the procedure, the pressure reading will tell you whether the problem lies in the pump or if it can be attributed to some other source, such as a leaky relief valve or a leak in the reservoir. On most H and M models (except the MTA) for example, the oil pressure reading should be 750 pounds or higher if the pump is in good condition.

If the pressure is lower than the proscribed reading, the problem can generally be attributed to one of six problems:

1. Low oil level in the reservoir
2. Excessive clearance in the pump
3. Relief valve is leaking
4. A leak in the reservoir (high pressure passage)
5. Oil is leaking past bushings on the drive shaft or oil seal
6. The slot in the back plate is too long, allowing oil to by-pass the pump gears

If the problem is traced back to the pump, overhaul is generally limited to disassembly, cleaning, and installing a seal and gasket package. If more extensive repair is needed, you'll most likely need to find a rebuilt pump or a salvage unit. A quality repair manual for your model will provide you with the pump part wear tolerances and the seal replacement procedures.

One word of caution—if the system needs to be flushed, don't use kerosene. Use IH Touch Control fluid or the recommended hydraulic fluid only.

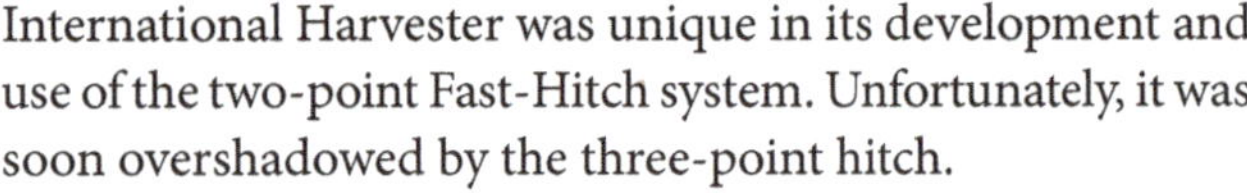

International Harvester was unique in its development and use of the two-point Fast-Hitch system. Unfortunately, it was soon overshadowed by the three-point hitch.

It's rare to find a hydraulic system on a tractor as old as this Farmall F-12. However, IH was one of the first to offer a self-contained system, optional on both the F-12 and F-14 to lift mounted implements.

Before disassembly of the reservoir and/or control valves, be sure to remove all accumulations of grease and dirt so you don't further contaminate the system.

At the very least, the hydraulic system on this Farmall model is going to need new seals.

It appears, from the evidence of torch welding on the back of the pump, that the previous owner of this Farmall Cub had problems with pump leakage.

Hydraulic Seals

It almost goes without saying that due to the high pressure within the system, no hydraulic circuit can operate without the proper seals to hold the fluid under pressure. Seals also serve the purpose of keeping dirt and water out of the system.

In general, hydraulic seals fall into one of two categories—static seals that seal fixed parts and dynamic seals that seal moving parts. Static seals include gaskets, O-rings, and packings used around valves, between fittings, and between pump sections. Dynamic seals, in contrast, include shaft and rod seals on hydraulic cylinder pistons and piston rods.

As for seal types, they can include O-rings, U- and V-packings, spring-loaded lip seals, cup and flange packings, mechanical seals, metallic seals, and compression packings and gaskets. Troubleshooting, naturally, consists of looking for leaks. However, even though the perfect seal should prevent all leakage, this is not always practical or desirable. In dynamic uses, for instance, a slight amount of leakage is needed to provide lubrication to moving parts.

On the other hand, internal leakage, either from static seals or excessive leakage from dynamic seals, is hard to detect. Often, excessive leakage from an internal seal must be indicated by other means, such as pressure testing.

As a general rule, it's usually best to replace all seals that are disturbed during repair of the hydraulic system—assuming they are available. As is the case with the engine, transmission, and most other major components, it's a lot cheaper to replace a few seals or gaskets during restoration than to come back later and do a repair job to correct leaks.

It may sound a little extreme, but you should also give seals the same care during handling and replacement as precision bearings. This means keeping them protected in their containers and storing them in a cool, dry place free of dirt until you're ready to use them.

Following installation, static O-rings used as gaskets should be tightened a second time after the unit has been warmed up and cycled a few times, to make sure they seal properly.

Dynamic O-rings, on the other hand, should be cycled or moved back and forth (as on a hydraulic cylinder) several times to allow the ring to rotate and assume a neutral position. In the process of rotating, the O-ring should allow a very small amount of fluid to pass. This is normal, since it permits a lubricating film of oil to pass between the O-ring and the shaft.

Unless directed otherwise in your service manual, you should always dip all cylinder and valve parts in Touch Control fluid when reassembling a cylinder and/or valve unit. It's also wise to lubricate all O-rings with Vaseline or an equivalent before installing them.

Although the power steering unit assembly on the steering shaft uses hydraulic pressure for the power steering system only, it should be inspected and repaired in the same manner as the main hydraulic valve and pump system.

Always check the integrity and condition of hydraulic hoses before reassembly.

Chapter 12

Electrical System

Unless you're restoring a late-model Farmall or McCormick-Deering tractor, there's not much to worry about on the electrical system. That's because there wasn't one—at least not as we think of it today. The closest thing to an electrical system was the magneto, which is essentially a generator, coil, and distributor wrapped up in one unit. Not until 1949 did International Harvester begin putting battery ignition and distributors on tractors as standard equipment. This, of course, called for a battery and charging system, as well.

Magneto Systems

Depending upon the year and model of your Farmall tractor, it could be equipped with any of several different magnetos. Early Regular models used an Edison-Splitdorf magneto until 1926, when the company switched over to an IH E4A. However, a Robert Bosch ZU-4 magneto was also available as special order. Later on, most models used IH's own F-4 or H-4 magneto.

Basically, a magneto works much like a miniature spring-driven generator. As the drive cog is rotated, the magneto drive wraps up a spring that is positioned between the drive link and the magneto rotor. At the appropriate moment, the spring is released and the rotor is quickly rotated within a magnetic field to generate a charge of electricity—hence, the clicking sound associated with magneto operation. The distributor portion of the magneto determines which of the cylinders receives the spark.

Due to the complexity of the spring-drive system, the need for special tools and testing equipment and the internal workings of a magneto, it's best to have any adjustment or rebuilding done by a professional shop; there are several listed in the appendix in the back of this book. However, there are some things you can check and replace yourself, including the points, condenser, rotor cap, and rotor tower cap.

You'll also need to make sure the magneto is reinstalled properly. This is especially important, since the magneto must initiate the spark before the piston reaches the top of the compression stroke.

On the other hand, the impulse coupling is designed to retard the timing during slow engine revolution, such as when it is being started. The result is a stronger spark for starting the engine and a reduced chance of kick-back, which can occur when the spark reaches the cylinder before the piston has reached the top of the cylinder. If you do not hear the click, your magneto may have a broken impulse coupler. Options for fixing this are to buy a new magneto or send the magneto to one of the businesses that specialize in restoration.

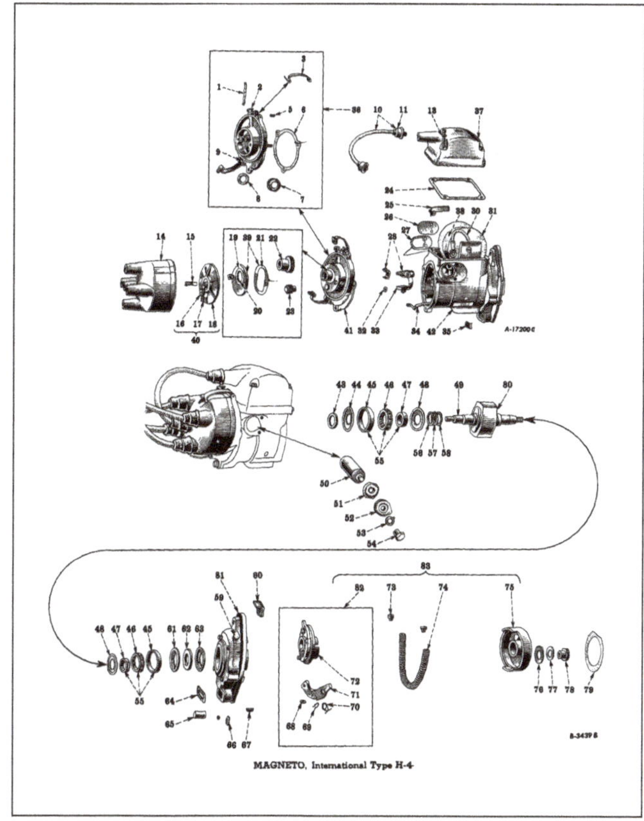

Due to the more complex nature of a magneto, like this International Type H-4 model, some restorers choose to send the unit out for professional restoration.

Since they required no electrical power, a magneto was used on almost all early model tractors to provide spark to the cylinders. Early models with their brass-clad magnetos are particularly appealing to collectors.

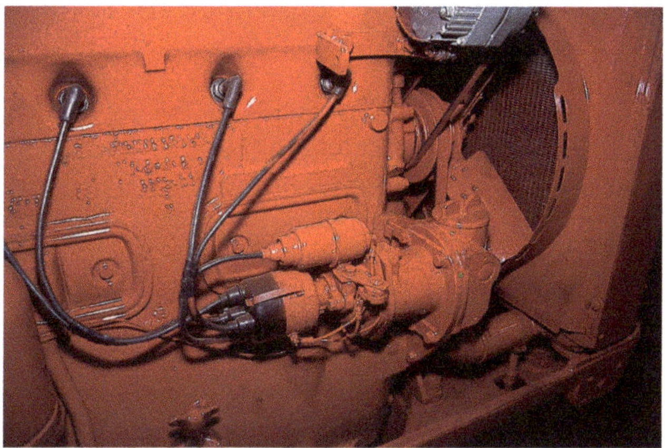

In the majority of cases, the magneto or distributor on IH tractors is driven directly off the governor shaft.

Since electrical lighting was still an option on this F-20, the battery box was hung on the side, almost as if it were an afterthought.

Magneto Inspection and Service

While there are a couple of ways to check for electrical output, one of the easiest for a novice is to attach a spark plug, via a piece of electrical wiring, to the coil output terminal. Then ground the spark plug to the base of the magneto and test for a spark while rotating the driven lug. A word of warning, though. Keep your hand clear of the coil and coil output terminal. If it's working properly the coil can put out nearly 30,000 volts, which you will obviously feel.

Most auto parts stores carry a spark tester that makes testing even easier. Basically, a spark tester looks like a spark plug with a large alligator clip. Hook the device to a plug wire and connect the clip to ground. When the tractor is cranked, the tester will flash if you have sufficient spark.

The presence of a spark won't be a guarantee that the magneto is putting out enough current, but it can give you an idea of how well the unit is working and if the spark is hot enough for ignition. If there isn't any spark, you at least know you're wasting your energy. The other testing alternative is connecting a special

Unlike modern electrical systems, the magneto operates like a generator, coil, and distributor wrapped up in one unit.

Just as it is with a distributor overhaul, it is a good idea to replace the magneto rotor tower and rotor if the parts are still available.

multimeter that records and stores readings in excess of 30,000 volts.

If you do choose to disassemble the magneto and try to service it on your own, the first thing you should do is inspect all the gaskets and insulators that isolate the generating components from the body of the unit. This especially applies to any bolts that protrude through the magneto body. Any voltage leakage can ground out the unit and make it ineffective.

You'll also want to clean all dirt and grime out of the unit, using a combination of compressed air and electrical parts cleaner. Finally, inspect and replace, if necessary, any seals or bearings that appear to be faulty. While you have the unit apart, it's also a good idea to have someone recharge the large horseshoe-shaped magnet that generates the magnetic field around the armature.

Before putting everything back together, apply a light coat of oil to all drive parts and bushings. Be careful not to over-apply the oil, though. Lastly, replace any tune-up components for which you can find parts. This includes the condenser, points, rotor tower, and rotor cap. It's easier and cheaper to replace them now than struggle with problems later on.

By carefully turning the pawl on the drive end of the magneto, you can listen for the click that indicates a spark is being created. By attaching a wire between the spark plug wire outlet and the case, you can also test for a spark.

Any overhaul of a magneto should include replacement of the points and condenser.

Timing the Magneto

There are two phases to magneto timing. The first step is to get the magneto roughly timed. The second is to get it adjusted for final timing.

To begin the timing process, you'll need to crank the engine until the No. 1 piston is coming up on the compression stroke and the ignition timing marks are in register. The marks will be in different locations, depending upon the model, but your service manual should explain this in detail. On Farmall A, B, and C tractors, for example, the flywheel mark "DC/1-4" should be in register with an indicator on the clutch housing cover as viewed through a hand hole in the bottom of the clutch housing.

Now, turn the driving lug on the magneto until the rotor arm is in the No. 1 firing position. To do this, hold the magneto in the same upright position as when it is mounted on the tractor. Carefully install the magneto and gasket on the governor case, making sure the impulse coupling lugs engage the slots on the governor drive coupling.

As soon as you have tightened the mounting bolts finger tight, rotate the body of the magneto as far as possible in the direction of normal rotation. Now, slowly turn the engine through one complete revolution until you have the No. 1 piston at the top of its compression stroke again and the timing marks are aligned as before. Slowly rotate the magneto in the direction opposite of its normal rotation until you hear the pronounced click of the impulse coupler. This indicates that the magneto is right at the point where it will fire the No. 1 piston. Tighten up the bolts, attach the spark plug wires to the cap and try starting the engine.

If you are hand-cranking the engine, be sure you follow the hand-cranking safety precautions in your manual, since you are trying to start an untested engine; there are several factors beyond magneto timing that go into backfires. Or, you may have made a mistake. This isn't a time to be careless.

Final timing involves carefully rotating the magneto by hand until it has been correctly positioned and tightening the mounting bolts.

Distributor Inspection and Repair

Since the distributor doesn't actually produce electricity, as the magneto does, but only serves to distribute the spark to the appropriate cylinder, service and rebuilding is much easier. Unless you plan to replace the spark plug wires, start distributor restoration by grasping the spark plug and coil wires and gently twisting the boot as you remove them from the cap. It's also a good idea to label the wires with tape and a marker so they can be reinstalled in the correct position.

If you plan to do much work on the distributor or engine, you may want to go ahead and remove the whole distributor at this time and clamp it in a vise where it will be easier to work on. Getting it reinstalled and timed correctly isn't that much different than reinstalling a magneto. However, to save some time later, make a note of which plug wire tower serves the No. 1 cylinder and make a mark on the side of the distributor housing that corresponds with this tower.

Now, pull the cap off and note where the rotor firing prong is positioned. Slowly rotate the engine so the rotor prong lines up with the mark on the house. Note, too, which direction the rotor turns. With the engine timed to No. 1, you can now remove the mounting clamps that hold the distributor on the drive housing. You'll note that the rotor shaft will rotate slightly as you remove the distributor due to the taper on the drive gear. You'll just need to remember to compensate when you reinstall it later. It's a good idea to reconfirm the proper timing anyway.

The first step in inspection and restoration is to check the inside and outside of the distributor cap and individual cap towers for cracks, burned spots, and corrosion. If there is any doubt about the condition, it's best to just replace the cap.

Before you throw the cap in the trash, though, check for carbon tracking around the plug towers and around the cap base. Defective spark plugs or spark plug wires can cause sparks to travel from the tower to the nearest ground, which is usually the mounting clip. The evidence is a small carbon trail that resembles a tiny tree root.

Rebuilding a distributor is not nearly as difficult as overhauling a magneto, since a distributor doesn't generate its own electricity.

A quality distributor rebuild should include new points, condenser, rotor, and spark plugs.

Check the inside and outside of the distributor cap and individual cap towers for cracks, burned spots and corrosion. If there is any doubt about the condition, replace the cap.

You should also take a look at the inside of the cap for carbon tracking that can indicate past problems with cross-firing, backfiring, or missing. In this case, the lines will usually connect from one post to the next.

Once you have inspected and cleaned all parts at the top end of the distributor, it's time to take a look at the bottom end. Start by checking for any free play or wobble in the drive shaft. Generally, there are two bushings or thrust washers on the shaft: one at the top and one at the bottom. If either one is badly worn, it should be replaced as part of the rebuild. To access the bushings, remove the drive gear, which is usually held on by a pin through the shaft and gear. This will allow the entire shaft to slide out of the housing.

Next, remove the points and condenser. Beneath the mounting plate you'll find a spark advance system of springs and counterweights. Replace any springs that are broken or weak and make sure the weights aren't rusted and that they move freely. The springs must be replaced in pairs and by springs of identical size. In the meantime, clean all pieces in solvent before reassembling.

Although some restorers like to reuse the points, touching them up with a file before they're reinstalled, a quality rebuild should include new points, condenser, rotor, cap, and spark plugs.

Setting the Point Gap and Distributor Timing

You will need to adjust the gap between the points. To do this, rotate the shaft to a point where the cam lobe separates the points, creating a gap. Now, using a wire-type feeler gauge, adjust the points to match the specifications in your service manual and tighten the mounting screws.

Be sure you remove any oil film from the feeler gauge before inserting it between the points. Oil on the contact points can cause the points to burn or become pitted. Finish off point adjustment by lubricating the rubbing block with a small amount of high-temperature grease.

At this point, you can install a new rotor, line up the casing marks, and reinstall the distributor in the housing. If you've compensated for any gear taper during installation, the mark on the distributor casing should line up with the rotor tang. If not, it shouldn't be off by more than one tooth.

If the distributor is completely off, though, there's no need to panic. Most service manuals include a detailed procedure for timing the engine or you can follow the general procedure which follows.

Start by cranking the engine until the No. 1 piston is coming up on the compression stroke, just as you would when timing a magneto. Continue turning the engine until the ignition timing marks are in register. Again, the location of these marks will vary depending upon the model you're restoring. In many cases, it will simply be a matter of lining up a notch on the crankshaft pulley with a pointer on the crankcase front cover.

Now, turn the drive shaft on the distributor until the rotor lines up with the terminal for the No. 1 spark plug. Hopefully, you marked the distributor casing earlier, making this an easy task. Offset the rotor a little to compensate for the gear mesh and slide it in, making sure the lugs on the distributor drive engage the slots in the drive coupling. Don't tighten the bolts yet, though.

Install a new distributor cap and spark plug wires, and finish the timing process with the ignition on. To do so, slowly turn the distributor in the direction of normal rotation, and watch for the exact moment that a spark occurs at the plug. If you missed it, back the distributor up and try it again. The engine should now be timed properly. Tighten the distributor mounting bolt(s). If you have one available, you may still want to check everything with a timing light.

Coils

Perhaps the easiest way to check the coil is to gently pull the wire that runs from the coil to the distributor and hold it about 1/8 inch from the engine block or a good ground. A strong spark should jump the gap when the engine is turned over. The coil should also be clean and dry. If you're not confident that it is in good condition, consider replacing it.

A refurbished and polished distributor like this one can really set off a restoration.

As was the case with the magneto, you'll need to make sure final adjustments have been made before tightening the brackets that hold the distributor in place.

Note the cylinder firing order molded into the engine block. This can help if there are any problems reinstalling the magneto or distributor, timing the engine, or reconnecting spark plug wires.

Generators and Voltage Regulators

If you ever did any experiments with electricity or with a generator in high school science or physics class, you may already have an idea how the generator on your tractor works. But if not, don't despair. It's not that complicated.

By the simplest explanation, one way electricity can be created is by moving a conductor through a magnetic field. So if you look at this principle in terms of a generator, the armature, which serves as the conductor, is moved, or in this case spun, inside of two or more magnets. But think back again to science class. Remember the time you wrapped electrical wire around a nail and attached it to a battery? You created an electromagnet. The generator on your tractor just uses a larger version.

So now you can envision the field coils as electromagnets attached to the generator case. As the armature spins within this magnetic field, electricity moves through the armature to where it is allowed to flow though the brushes.

The wire that makes up the coils actually begins at the F terminal of the generator, winds its way around the case, and terminates either at a third brush on a three-brush generator or connects to the wire going to the output brush or A terminal on a two-brush generator. Unless the generator has been replaced, it most likely uses two coils.

The magnets are actually a two-piece arrangement consisting of a coil of wire that fits around a pole shoe made of a special kind of metal. As the armature spins within the magnetic field, it begins to generate electricity—some of which is used to charge the battery or run electric lights, while the rest goes back into the field coils to make the magnetic field even stronger. The role of the regulator is to control the generator by manipulating the ground connection of the field coils.

Among the most important components are the brushes, which serve to gather the electricity that is being produced. The brushes ride on the commutator and allow the generated electricity to travel to the voltage regulator, and ultimately the battery or other load, then back into the field coils.

As a result, the primary problem areas on a generator are the brushes, commutator, and bushings. Generally, brushes should be replaced if they are worn more than halfway. Quite often the bushings and bearings that support the main shaft will also be worn and require replacement.

To perform any of these repairs, however, it will be necessary to disassemble, clean, and inspect the generator. Don't try to remove the field coils unless it is absolutely necessary. If you take them out, it will be difficult to get them back in without the proper tools. There are only two things that can go wrong with the field coils anyway—either the wires have lost their insulation somewhere and are touching ground (such as the generator casing), or they have broken and have created an open circuit.

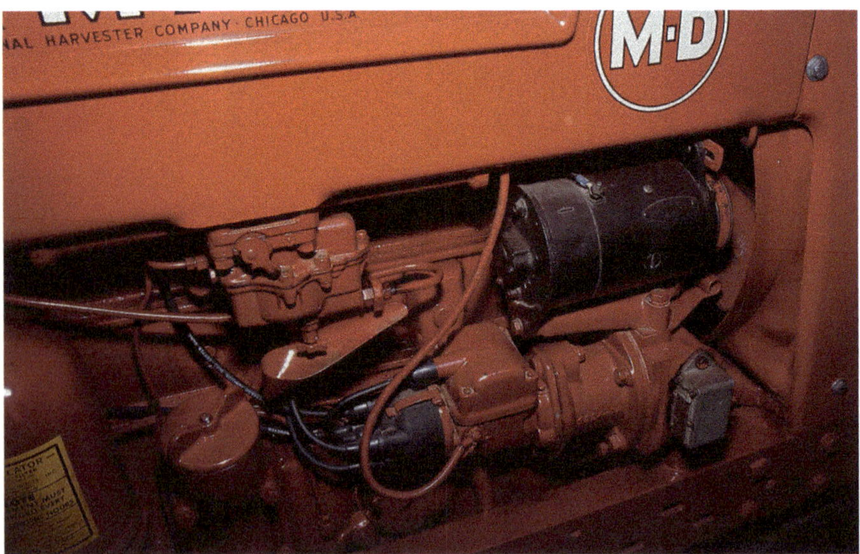

The function of the generator is to replace any electricity in the battery that has been used elsewhere in the electrical circuit. Hence, generators weren't needed until the advent of lights and electric starters on tractors.

The first check of a generator or starter is turning the shaft to check for bearing play or stiffness.

Considering the cost and availability of rebuilt generators, most restorers will tell you it's seldom worth your time to try to repair a generator yourself. Moreover, most communities have a machine shop or automotive shop that can test and rebuild your generator to factory specifications.

The same could be said about the voltage regulator. If there is doubt about whether it is working properly, your best bet is to take it to a shop that specializes in starter and generator rebuilds and let them test it. If it can be repaired, they'll be able to take care of it, and if not, they should be able to suggest the correct replacement.

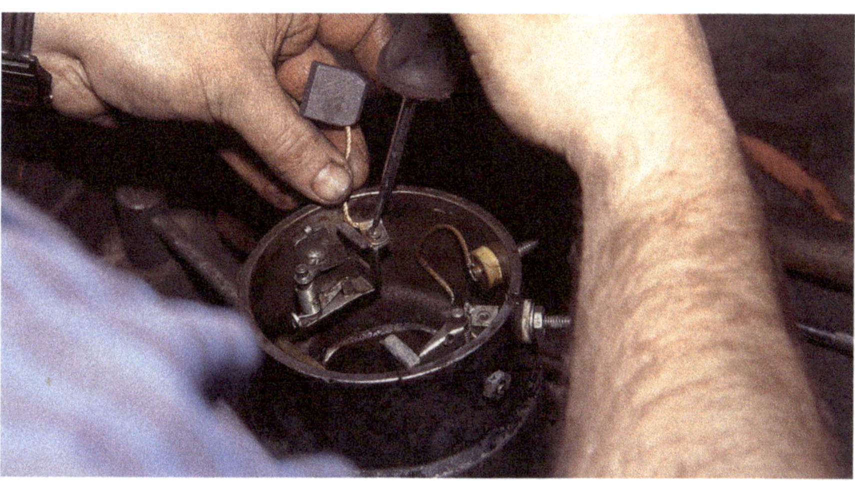

The brushes on generators and starters alike should be replaced if they are worn more than half-way.

Whether you're overhauling a starter or generator, it's worth the time and money to put in new bearings.

Starters

Not every Farmall or McCormick-Deering tractor is going to have a starter. Even when electric starters became available, they were only an option on many models.

In essence, the starter is much like the generator, only it operates in the opposite manner. Instead of generating electricity, it takes electricity and uses it to turn a drive sprocket. However, like the generator, it contains an armature, coils, brushes, and commutator that can wear or short out in much the same manner. Hence, overhaul consists of inspecting the brushes for good contact with the commutator and making sure the latter is reasonably clean and smooth. If it is not, it will need to be turned down on a lathe.

Just as you did with the generator, you'll also need to check for worn, dirty, or damaged bearings. Again, it may be easier and less costly in the long run to have these things done by a shop that specializes in starter and generator repairs or trade it for a rebuilt unit.

Before you pull the starter off the tractor, though, you need to realize that some tractors have a neutral start interlock switch that prevents the tractor from being started if the tractor is in gear. If the interlock switch is faulty, you may be wasting your time on the starter itself.

You may also have problems related to the battery or loose or corroded connections. To check for these, connect a fully charged battery to the starter using a set of jumper cables. If there is a significant difference in the way the starter turns over, check the battery; then inspect, clean, and tighten all starter relay connections as well as the battery ground on the frame and engine. If the starter still does not crank with the jumper cables, plan on removing and replacing or repairing the starter.

The starters on early Farmall and McCormick-Deering models were pretty basic. In most cases, it's mounted right behind the engine, where it directly engages the flywheel.

Wiring

Unless your tractor has been treated with tender loving care for the past fifty to seventy-five years, it's doubtful that you will get by without replacing at least part of the wiring. At the very least, you will want to replace the spark plug wires as part of the engine rebuild.

However, replacing crimped, spliced, and inferior wiring on the rest of the tractor not only improves the looks of your restored tractor, but it also can be a safety measure. Wiring with cracked insulation and wires on which the old, cotton braiding has dry rotted can ground out electrical components—and at the worst, create a fire hazard.

The first step in wiring restoration is locating a wiring diagram. Hopefully, this will be included in your service and repair manual. If not, you'll have to trace the wiring from the power source, or the battery, to each switch and component. Don't put too much stock in what you find, though. Previous owners may have replaced the original wiring with a different gauge of wire, the incorrect type of wire, or they may have even taken shortcuts with the routing.

If there is any doubt, talk to other tractor owners or try to find a well-restored model like your own and take notes. As a general rule, you should use at least 10-gauge wiring for circuits that carry a heavy load, such as from the generator. Switches and other components can be wired with 14-gauge wire. Remember, the larger the gauge number, the smaller the wire diameter.

T. W. Cook, a Farmall H enthusiast from Georgetown, Texas, insists a 6-volt system is adequate if the cables to and from the battery and starter are 2 gauge or larger in size and in good condition.

If you intend to use your restored tractor strictly as a work tractor, you might want to install an alternator and convert the electrical system to 12 volts. On the other hand, that may have already been done by a previous owner and you'll need to find a generator if you want to restore it to original.

T. W. Cook, Farmall H enthusiast from Georgetown, Texas, says he often gets asked about converting a Farmall tractor to a 12-volt electrical system. "My answer is, 'Why?'" he says. "These tractors were designed to start perfectly well on six volts and will today if you have proper cables and a good battery."

Cook says the first thing to check is the battery cables. He insists that many tractors he's seen have had the cables replaced with thin 12-volt cables. "If you have these," he says, "you need to throw them away and replace them with big, two-gauge or better cables. Make sure your ground cable—remember, the H has a positive-ground system—goes directly to one of the starter mounting bolts and makes good connections on both ends. Then make sure the cable from the battery to the starter switch and on to the starter are in good condition."

Cook advises that if you can't find proper replacement cables and all else fails, get some two-gauge welding cable and make your own. "It's not that hard," he relates.

If you have a lot of wiring to replace, it might be easier just to purchase a complete wiring harness for your tractor. Available through various sources, an appropriate wiring harness is made up with the correct gauge and color of wiring for each switch, gauge, and component and is pre-wrapped to match the routing.

Finally, you'll need to consider what role historical accuracy plays in your restoration goals. If you're restoring the tractor as a working machine, you can get by with modern automotive wiring in the correct gauge and crimp-style connections. However, if you're going for an accurate resto-ration, you'll need to locate the appropriate gauge of lacquer-coated, cotton-braided wiring for any tractor that used cotton-covered wire as original equipment. Generally, cotton-braided wiring is appropriate for any tractor built before World War II.

The use of lacquer-coated cotton braided wiring—which was common on tractors built before the mid 1950s—really sets off an early F Series restoration.

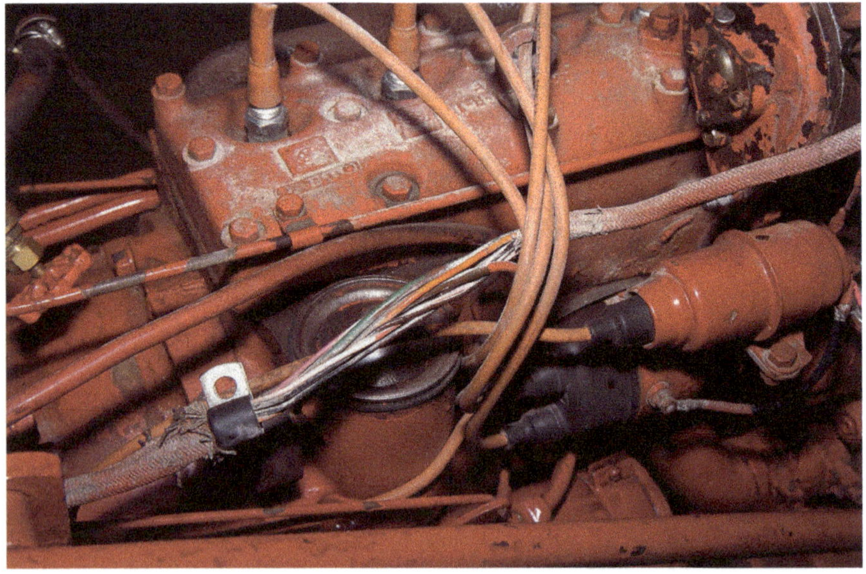

Sometimes it's easier to buy a complete wiring harness than to try to figure out how to rewire or repair worn and patched wiring like this.

Spark Plug Wires

Contrary to their rugged appearance, spark plug wires are quite sensitive. That's because most secondary wires consist of a soft-copper-core wire surrounded by stainless steel or carbon-impregnated thread mixed with an elastomer-type conductor. The outside covering of a heavy layer of insulation prevents the 12,000 to 25,000 volts from bleeding out when the wire is carrying current.

Consequently, the resistance-type wires do not handle sharp bending or jerking and can break internally, ruining the wire. Excessive exposure to oil and antifreeze can chemically break down the coating, as well. Perhaps the most damaging effect is caused when someone pulls on the wires to remove them from the spark plugs. This can separate the conducting material, causing internal arcing.

Therefore, it's important that you keep spark plug wires clean and separated from each other with any wire clips that are integral to the routing. To remove wires from the plugs during service or restoration, grasp the rubber boot, not the wires. Keep in mind, too, that secondary wires, including the coil wire and spark plug wires, can appear in good condition, yet be faulty.

One way to check their condition is to measure the wire resistance using an ohmmeter. In general, the wires should register around 8,000 to 12,000 ohms per foot. This will also test the wire for continuity, ensuring that there are no breaks in the copper-wire core.

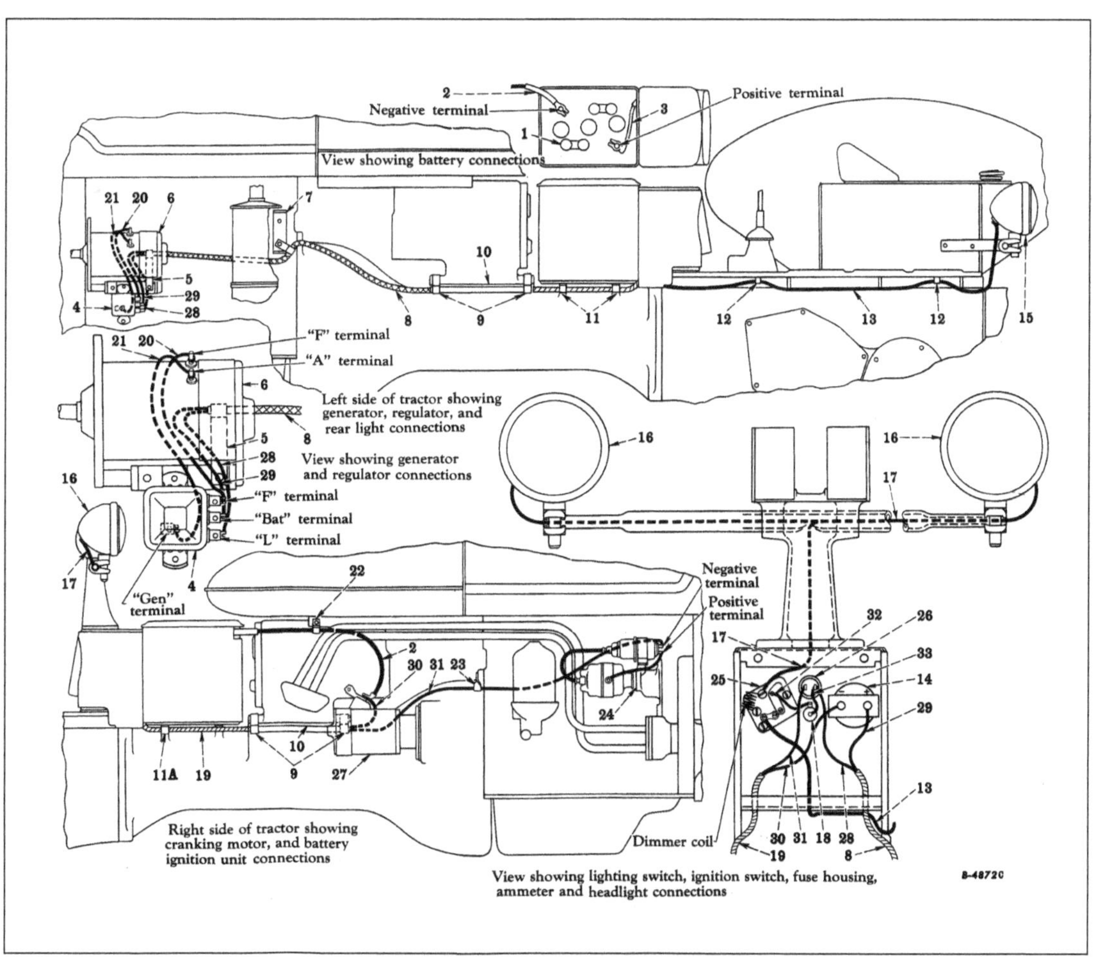

A wiring diagram can be a tremendous help when rewiring a tractor or tracing electrical problems.

Lights

Depending upon the age of your tractor, you may not even have to concern yourself with refurbishing the lights. Electric lighting systems were available on IH tractors as early as the 1930s, but had to be ordered as an option. Lights on tractors didn't become standard equipment until the late 1930s.

Besides broken lenses and deteriorated wiring, the most common problem you're likely to encounter is a rusty, faded, or worn reflector. Most restorers start the renewal process by disassembling the light, then cleaning it inside and out. One restorer likes to bead blast the light housing to a smooth finish—although he admits paint stripper can have the same effect.

Then, it's simply a matter of using a galvanizing-effect paint to spray paint the reflector. The outer shell can be painted at the same time you're painting other tractor components. Should you have any trouble finding a replacement gasket for reassembly, you might also want to use a tip provided by Jeff Gravert, a tractor restorer from Central City, Nebraska. He says a good substitute for the gasket that fits between the lens and the reflector is a strip of caulking that comes in rolls and pulls off like a piece of cord. The caulk also helps hold the lens in place.

Should you need to find a replacement light, swap meets, salvage yards, and dealer parts counters are all good sources. There are also a number of vendors that offer both reproduction and refurbished lights for sale.

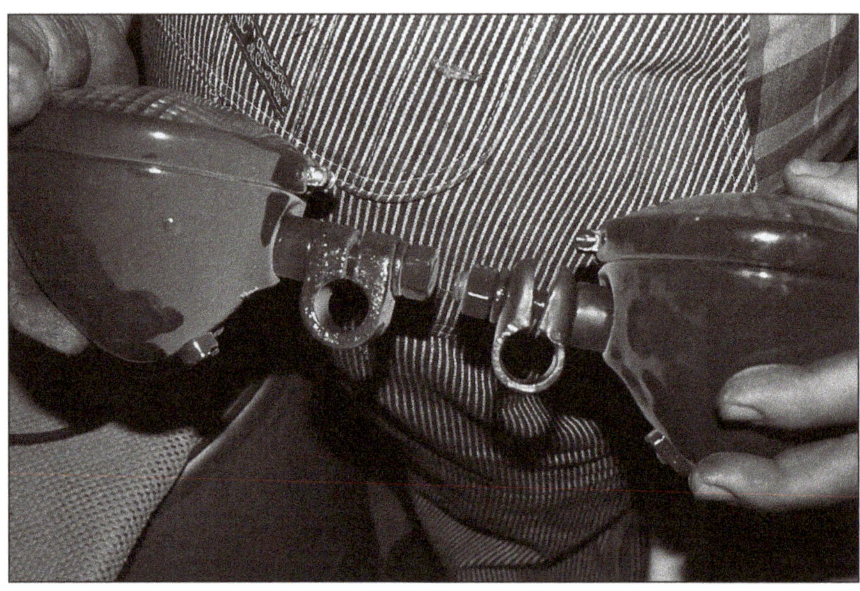

Walter Bieri discovered that IH used two different types of headlight brackets on their Farmall tractors: One type is cast and the other is stamped.

When he has spare time between restorations, Bieri often restores individual parts, like this lineup of headlights, so they're ready when he needs them.

In the early 1950s, IH switched from bulb-type headlights to sealed beam units. These were correct for this 450, but you'll see bulb-type lamps inaccurately used on some restorations, simply because the sealed beam units are expensive and hard to find.

Gauges

Gauges can be a real problem for the restorer trying to obtain an authentic look. Often the gauges work fine, but look a little worse for wear, detracting from the whole appearance of a nicely restored model. If this is the case, you have a few different options. Stewart-Warner (S-W), for example, makes gauges that work fine for many older tractors. Many of them even look a lot like the original except they have the S-W logo at the bottom of the gauge face. However, if you're not terribly concerned about authenticity, you can replace any gauge that is no longer working with an off-the-shelf model.

On the other hand, if you're going for authenti-city, you might want to contact one of the vendors listed in the back of this book and find a specially built reproduction. Don't overlook the local dealer, though. He may just have some new old stock (NOS) parts that were made for your model.

If a gauge is not available from any other source, however, there's still the salvage yard. Depending on the reason the tractor ended up there, it may have a gauge that's in better shape than the one on your tractor. If the face plate is in relatively good shape, it's easy enough to refurbish the rest of the gauge.

First, you'll need to carefully remove the bezel ring that holds the glass in place. On some gauges, you may have to bend up the lip around the edge of the gauge to do so. Now, it's just a matter of cleaning it up, making sure the mechanisms work properly, and repainting it.

The appropriate parts are arranged for a sealed-beam light restoration. Bieri says he often substitutes a cheaper, off-the-shelf sealed-beam in place of the higher-priced version from the Case IH dealer.

New gauges, complete with the IH logo, really set off the dash on this Model 450 restoration.

Chapter 13

Fuel System

There can be a wide variation in the amount of work a fuel system is going to need, depending upon the age of the tractor and whether it was running at the time you bought it. If you had a chance to drive it before the purchase or before you started tearing it down, you should have a good idea about how smoothly it was running.

On the other hand, if you're restoring a treasure that has been sitting in the weeds or an old barn for the last twenty years, chances are there's a lot of rust, varnish, water, and who knows what else in the system.

Like most tractors, the first Farmall and McCormick-Deering models were all-fuel only. In fact, the F-30 was never offered in anything but an all-fuel model. However, by the time the letter series made their debut, customers could choose from gasoline, all-fuel, liquefied petroleum gas (LPG), or diesel-powered models.

Fuel Tank

You can rebuild the carburetor, clean or replace the fuel lines, and change the fuel filter, but all that does little good if the fuel tank was the source of contamination. So the first step in fuel system overhaul should be to clean the tank and reseal the interior if necessary.

Everyone seems to have their own story about how to clean a fuel tank. Some have been known to stick the sandblast nozzle in the fill opening and move it around to hit all sides with silica sand or glass beads. The risk, of course, is that you might just blow a hole through any weak spot in the tank. Plus, you'll need to get all the sand out of the tank.

A more common option is to fill the tank about one-third full of water and add a few handfuls of 1/2-inch nuts, shingle nails, or pebbles. The key is to provide a slight abrasive action to clean up the inside. A word of caution is in order, though, particularly if you want to avoid frustrations later on. Check to see if the filler neck extends into the tank to act as a baffle that keeps fuel from splashing back out. If it does and you add a non-metallic abrasive like pebbles or even brass nuts, you may have a hard time getting all the material out of the tank. It will be a little like trying to shake pennies out of the slot in a piggy bank. Most professional restorers prefer nuts and bolts, because the stragglers can be fished out with a magnet.

The next step is to agitate the tank rather vigorously with this mixture sealed inside. One restorer claims the best way to agitate the mixture, assuming you do some farming, is to strap the tank to a tractor wheel with bungee cords and let it rotate while you do a day's field work. Another restorer says he does the same thing, but simply blocks the front wheels on a tractor, locks the brakes, and jacks up one rear wheel to which the tank is strapped. Then, he lets the tractor idle in gear for four or five hours, letting the tractor wheel work much like a rock tumbler. Yet another restorer says he secures the tank in the back of his pickup and hauls it around for about a month while he's working on other parts of the tractor. If you're going to use that method, it helps if you live on country roads. Anyway, you get the picture. You need to agitate the tank vigorously enough and long enough to scour all the rust and residue out of the tank.

If the tank is in really bad shape, you might want to start by adding a lye-based cleaner to the initial mixture for the first fifteen or twenty minutes and then switching to a clear water-and-abrasive mixture. Once the tank has agitated for a sufficient period of time, remove the abrasive material (pebbles, nails, or nuts) and rinse the tank with clean water. You may have to repeat the rinsing process several times until you get clean water coming out of the tank.

If you find that the fuel tank leaks, do not try to solder it yourself. Regardless of what kind of instructions your friends have given you—like filling the tank with exhaust gas from your car's exhaust pipe, which supposedly makes it safer—it is impossible to get the fuel tank clean enough to solder safely in a home shop. Shops that repair automobile gasoline tanks usually steam clean the insides for an hour or more to ensure that no residual gasoline is emitted from the pores in the metal during heating. Even then, soldering a gas tank can be a dangerous proposition. That's why many professionals also fill the tank with an inert gas or liquid before heating the tank.

One thing you can try on your own is patching the hole with an epoxy, assuming the patch will be hidden beneath the tractor sheet metal. Several restorers have reported success with materials marketed as Magic Metal, J-B Weld, and other "gas tank menders" sold in

Distillate or kerosene engine tractors always included a small gasoline tank for starting the engine. Ironically, the gasoline tank was mounted in a variety of locations, including within the distillate tank, inset in the tank or mounted next to the tank, as was the case with later models on which distillate was a seldom-ordered option.

automotive stores. The key is getting the surface clean with a good parts cleaner prior to mixing and applying the epoxy.

Once the tank has been repaired or proven to be free of leaks, it still makes good sense to coat the interior with a fuel tank sealer. One restorer, who shall remain nameless, got a tractor completely back together and painted, only to put five gallons of gasoline in the tank and discover a leak in the fuel system. He eventually traced it to a few pinholes where the fuel tank contacts the support straps that hold it in place.

Most sealer formulas recommend that you first etch the tank with phosphoric acid or an acid metal-prep solution to stabilize any remaining rust prior to adding the sealer. Be sure to leave the fuel tank lid off, though, when rinsing the tank with acid, since the reaction with the metal creates a gas. Then rinse the tank several times with clean water and air dry the tank with a warm air source to prevent any further rust.

As soon as the tank interior has adequately dried, pour in enough sealer to cover all sides of the tank interior. In most cases, the sealer instructions will tell you to allow several days for the material to cure before adding fuel. While you're waiting, you can finish up the fuel delivery system by cleaning the sediment bowl assembly and replacing all gaskets and screens.

A vital step in fuel system overhaul is cleaning the tank and, if necessary, resealing the interior with a quality tank sealer.

Fuel Hose Inspection

The fuel lines and fuel filter may look fine right now, but taking a few extra minutes to thoroughly examine them could still save you a lot of headaches later. Start by inspecting rubber fuel hoses for kinking, pinching at tight bends, and internal swelling. Also, make sure fuel lines are not running near an exhaust manifold or pipe, which could lead to a vaporizing problem on hot days. If the fuel turns into a gas in the line, it can cause the fuel circuit to "vapor lock" and stop delivering fuel to the carburetor.

Finally, make sure fuel is flowing freely to and from the fuel filter. A partially plugged fuel filter can lead to a leaner fuel mixture and cause backfiring, spitting, and misfiring. Unfortunately, when the engine dies, back pressure from expanding vapor can push debris from the filter back into the fuel tank, thereby hiding the problem. The engine will start and run like normal until debris, once again, finds its way back into the filter element.

Carburetor Repair

To your benefit as a tractor restorer, the carburetor on most vintage farm tractors is not as complex as it would appear. To begin with, there was no such thing as a fuel pump on most early tractors. The fuel tank was simply mounted above the engine and the fuel was fed to the carburetor by gravity. The carburetor itself is equally simple. International Harvester used a number of different carburetor brands and models over the years, depending upon the tractor model. They include Marvel-Schebler, Zenith, Carter, and International Harvester's own version.

Still, adjustments in most cases are limited to the idle-mixture adjusting needle and load adjustment screw. Clockwise rotation of the idle-adjustment needle reduces the amount of air and leans the mixture. Most models, except for the 3/4-inch updraft used on the Cub, are also equipped with a fuel controlling main jet or load adjusting screw, as well. Turning this screw clockwise leans the mixture under load.

Internally, about the only adjustment that is ever needed is to bend the float stem to change the fuel level in the bowl. The float setting is important because the fuel level in the bowl plays a critical role in low- and high-speed adjustment.

In principle, the float bowl acts as a reservoir to hold a supply of fuel for the carburetor. However, it's important that the fuel in the bowl remain at a consistent depth, since the fuel level regulates fuel flow to the carburetor itself. As fuel fills the bowl by gravity, the float raises on a hinge and pushes the needle valve into a seat to shut off the fuel flow. In effect, it works in much the same way as the float and valve in the toilet tank in your bathroom.

If the fuel level in the bowl is too low, the engine does not respond readily when accelerated and it will be difficult to maintain carburetor adjustments. If the fuel level in the bowl is too high, it can cause excessive fuel consumption and crankcase dilution. Plus, it can cause the carburetor to leak. Again, it will be difficult to maintain carburetor adjustments.

Depending upon the carburetor model used on your tractor, the top of the float should be in the range of 1 5/16 to 153/64 below the top of the bowl. The exact dimension for your tractor should be listed in your service manual.

Carburetor Rebuilding

The first step in rebuilding a carburetor is to remove it from the tractor and get it cleaned up. Start by closing the valve on the fuel tank, if this hasn't already been done or if the fuel tank hasn't already been removed, and disconnect the fuel line from the carburetor. You'll also need to disconnect any choke cables and governor linkage on most carburetors.

Now, remove the carburetor from the intake manifold and the air cleaner and move it to a clean workbench or other area where you can disassemble it without losing pieces. Remember that until you empty the bowl, the carburetor still contains a small amount of gasoline. So treat it as flammable until you've cleaned it out. If the carburetor is equipped with a drain plug or a drain valve, and if it's not rusted in place, it's best to drain the fuel before you go any further.

Begin by removing the top half of the carburetor and dumping any gasoline that is still in the bowl into a safe place. Next, carefully disassemble the carburetor, inspecting all parts for wear as you go. Before you remove the idle-mixture and main-jet-adjusting needles, though, carefully tighten each against the bottom of the seat, noting how many turns it takes to do so. When reassembling the carburetor, you can again tighten the needles to the seat and back off the recorded number of turns. That will at least give you a starting point to dialing in the carburetor. If you have a good service manual, it will also tell you how many turns open are needed for initial settings. Of course, final adjustment must be made when the engine has been warmed up and is running.

Be sure you make notes—including notes about the orientation of any gaskets you remove—if you have any doubts about how the carburetor goes back together. Many of the parts will be replaced while installing a carburetor rebuild kit, but don't throw anything away until you know you have the proper replacement part. Some kits are applicable to more than one carburetor,

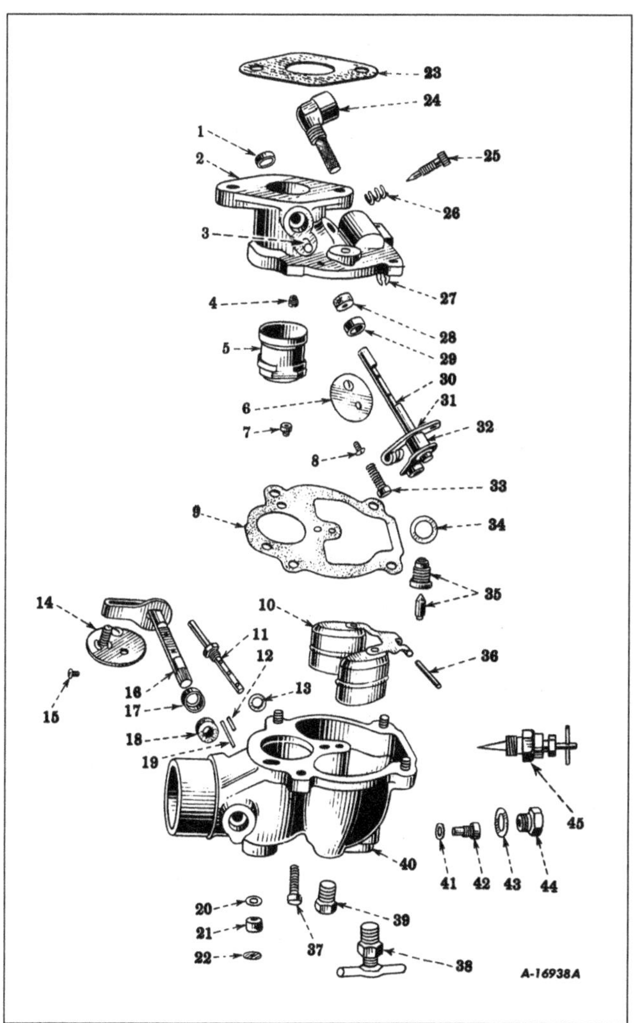

With the correct parts kit, a set of instructions and/or an exploded view drawing such as this one for a Zenith Model 161, even a novice restorer can perform minor carburetor repairs.

so there may be parts you don't need—and that means you must be able to match the parts you *do* need! Next, soak the two halves of the carburetor, along with the components you've removed, in a new container of carburetor cleaner for at least twelve hours or for the amount of time recommended on the label. Many carburetor cleaner solutions come with a parts bucket, so use it to turn and move the components from time to time.

Some restorers have also used a sandblasting cabinet and glass beads to scour the two halves of the unit once all the parts have been removed. You need to use care, though—and never use sand. Otherwise, you can quickly ruin the brass jets that remain in place.

You'll also need to clean these brass jets, either by removing them with a screwdriver or cleaning the passageway with a sturdy piece of nylon fishing line and an air hose, directing the air in the opposite direction as the fuel flow. Some jets are pressed into place and can't be removed without destroying them, which means you need to know ahead of time whether a replacement is available. If a jet is removable, make sure you have a screwdriver that fits the slot securely. Brass parts are easy to strip or damage.

The drip pan on this early model carburetor allowed the operator to restart the engine on gasoline in the event the engine died in the field. Because the engine would seldom restart on distillate, the operator had to drain the fuel into the pan, allow it to evaporate, and switch over to the gasoline tank.

To remove the carburetor, it's necessary to first remove the fuel line and all throttle and choke linkages. Then, unbolt it from the intake manifold.

Farm equipment dealerships and automotive parts stores are also a good source of carburetor kits. Notice that it even includes full instructions for carburetor overhaul.

Soaking the disassembled carburetor in a bucket of cleaner should be the first step in the overhaul process.

Carburetor Reassembly

Once the carburetor has been thoroughly cleaned, it's time to put it all back together and make the necessary adjustments. Hopefully, you were able to find a kit for your carburetor that contained all the appropriate parts.

It's important to inspect all moving parts and replace those that are damaged. Carefully inspect all adjusting screws, seats, inlet needles and cages for ridges, nicks or depressions that could affect the air/fuel mixture. If the throttle-shaft bushings or seals are worn to the point they are letting air leak into the carburetor, they're also going to affect the gas-air mixture. In some cases, the throttle shaft itself may have a groove worn into it. Some models have renewable throttle shaft bushings and others do not. On unbushed models, you'll have to compensate by replacing the throttle shaft and/or throttle body.

In the process of reassembling the carburetor, you also need to make sure all gaskets are properly oriented. Otherwise, you may block a vital orifice or passageway.

Also, when checking the float height, be sure the gasket has been positioned on the bowl half of the carburetor. The measurement specified in your service manual or kit instructions is almost always taken from the surface of the gasket to the bottom surface of the float.

Finally, when reinstalling the main-jet and idle-speed mixture screws, be sure to screw them all the way in and then back them out the number of turns recorded during disassembly, or as instructed in your service manual.

Most carburetors are not as complicated as they might appear. If you don't think you can handle installation of a carburetor kit, though, there are plenty of businesses that offer rebuilding services.

Be sure to clean all fuel passages, including the inlet and outlet from the sediment bowl.

Make sure all gaskets are oriented correctly when reassembling the carburetor. Otherwise, you may be blocking off a vital orifice or passageway.

Diesel Systems

International Harvester joined the diesel revolution as early as 1934, when the company introduced the WD-40. However, the company attained new leadership in the diesel tractor industry in 1941 with the introduction of the MD. Few tractors to that point were even using a diesel engine in a farm tractor, let alone a row-crop model. Surprisingly, the MD used the same 248-cubic-inch engine block used in the gasoline-model M. The diesel version, however, had a few differences. It used aluminum pistons and two additional main bearings. It also used a rather unusual starting system, compared to other diesel engines. While most other diesel engines, before and after, used a pony engine or heavy-duty starter to crank the diesel engine, IH designed a unique engine head that contained both spark plugs and diesel injection units. Hence, the engine was started on gasoline. Then, when it was warmed up sufficiently, the operator switched it over to diesel, using a lever on the platform.

Unlike gasoline, which is vaporized and ignited with a spark, diesel fuel is injected directly into the cylinder at pressures up to 2,500 psi. In fact, some of today's diesel engines utilize pumps producing up to 5,000 psi pressure at the injector. It is this high injection pressure, combined with cylinder compression, that creates the heat needed to ignite the fuel. This means that the injector pump needs to force the precise amount of fuel into each cylinder, via an injector, at exactly the right time. That's why the injection pump is generally geared to the crankshaft.

That's also the reason injection-pump testing and rebuilding is best left to a professional who has the knowledge and the equipment to work on it. One thing you can do, though, is make sure the injector is the correct size for the tractor model and engine you're rebuilding. Any overhaul or replacement, however, requires access to complete nozzle testing equipment. In the meantime, you need to remember that cleanliness is vital when working on diesel systems. It's also important that the fuel system is purged of air whenever the fuel filters have been removed or when the fuel lines have been disconnected. Again, the process for purging air on your tractor should be outlined in your service manual.

International Harvester tractors were innovative in several ways. One was that the first diesel engines started on gasoline and were switched over to diesel. If you were to look on the other side of this diesel engine, you'd find a distributor, coil, spark plugs, and all the other components of a gasoline engine.

Oil-Bath Air Filters

For many of us, an air filter is a square or round element composed of aluminum screen and folded, paperlike material that traps dirt particles as air flows to the carburetor. Once it gets dirty or has been in place a certain length of time, it's tossed and replaced with a new one.

While some vintage tractors may have that type of filter, it's more likely that the engine on your tractor has an oil-bath air filter. After all, replaceable air filters didn't become a feature on most farm tractors until the 1960s.

Let's first look at how the oil-bath air filter works. As you probably noticed, the filter itself looks like it is made out of a pile of metal shavings. What's more, there's a cup at the bottom filled with oil. When you start the engine, a certain amount of oil is sucked out of the oil cup or pan and onto the metal shavings, which then filter the air as it flows through the canister.

The filter canister is designed to be just the right height and size so that the engine can pull oil up into the screen along with the air, but without pulling it on into the carburetor and engine. If you notice, too, incoming air is drawn into the filter through a center pipe that leads to the bottom of the canister. As a result, any heavy dirt particles should fall directly into the oil cup. Lighter particles, of course, should be trapped on the oil-soaked filter surface as the air moves upward through the outer portion of the canister toward the carburetor.

The breather and oil-bath air filter are an important aspect of the fuel system. If they're clogged, there won't be enough clean air to mix with the fuel in the carburetor and no amount of adjustment can cure the problem.

To work correctly, the oil-bath air cleaner cup must be filled to the recommended level with the correct weight of oil.

Be sure to replace any damaged hoses that connect the carburetor to the air intake. Air leaks can negatively affect the air to fuel mixture.

Now that you understand how the filter works, it should also be easier to visualize the potential problems. The first comes with using the wrong weight oil. If you add oil that is too light, it can be drawn beyond the filter and into the engine. Using oil that is too heavy will have the opposite effect—not enough oil will be drawn up into the filter element, and much of the air-cleaning surface will go unused.

As inconvenient as it may sound, oil-bath air filters were designed to be cleaned and refilled daily when in use. In really dusty conditions, a farmer sometimes had to service the air cleaner a couple times a day. Naturally, the air cleaner is going to work best when the oil level is at the recommended level. However, letting the oil cup fill up with sludge can be even more detrimental. Simply adding more oil to the cup, in fact, can make it worse. When the particles-to-oil ratio gets to a certain level, the dirt will begin to hang onto the cleaning surfaces. Eventually, instead of just clean air being sucked into the intake, chunks of dirt and sludge are going with it. So it's important to dump the old oil and wipe the oil cup out on a regular basis.

Finally, you'll recall that the air cleaner was designed to be just the right size to match engine air intake. Otherwise, too much or too little oil is drawn into the filter canister. That means that any replacement air filter needs to be similar in size and design, if not identical to the original. By the same token, if you make dramatic changes in the engine that are going to affect air intake, you will need to make comparable changes in the air-cleaning system.

When it comes to repairing an oil-bath air cleaner, your biggest enemy will likely be rust, particularly if the oil in the bottom pan has long been replaced with water. Although some restorers have had success rebuilding small holes with epoxy or J-B Weld, about the only option when you're faced with a rusted-out canister is to locate a replacement.

Manifold Inspection and Repair

Just as an air leak in the carburetor can affect how smoothly your tractor runs, so can a crack in the intake manifold. When that is the case, one option is to have a local welder or machine shop make the repair. Unfortunately, most intake and exhaust manifolds are made out of one of four types of cast-iron material—white, gray, malleable, or ductile iron. To weld it, the material must be properly prepared, preheated, and welded with the appropriate method. However, the type of material must first be identified, which involves one or more of the following tests: chemical analysis, a grinding test that identifies the types of sparks a grinding wheel gives off when in contact with the material, and a ring test that helps identify the material by the type of ringing sound it gives off when struck with a hammer.

Should you decide to try arc welding a cracked manifold yourself, it's important to use a high-content nickel/cast rod or a nickel/cadmium rod with a cast-iron-friendly flux. Also, try to preheat the manifold with a torch. If you try to lay a long bead of weld on a cold manifold, it could easily warp and cause a sudden stress crack somewhere else. If preheating is not possible, strike an arc and weld only an inch or so of the crack. Then stop and let the heat spread to other parts of the material.

One alternative to arc welding is brazing the manifold using a brass rod melted into a prepared groove on the manifold crack. Start by locating the crack and grinding a groove along its length with a grinder. Extend the groove a half inch or so beyond the crack. Then use a coarse file to remove the grinder marks from the groove. This helps remove any graphite particles that could prevent the brazing material from adhering to the iron.

As for the brazing material, it's best to select a brass rod that is high in copper content with some nickel added. Also, select a torch tip that has a high heat output with low gas pressure. As with arc welding, it's helpful to preheat the material to be welded so it won't crack under isolated heat stress. Once brazing has been completed, try to cool the manifold slowly, using a bed of sand, if available.

One last option when trying to repair a manifold is to use an epoxy, such as J-B Weld. This is a particularly viable option on an intake manifold that is in a relatively cool area of the engine and when the crack is not in a stress area, such as the areas around the mounting flanges. Just make sure the area is clean, free of grease and grit, and prepared according to the directions on the epoxy package.

Of course, the easiest alternative is to just locate a replacement manifold at a salvage yard, flea market, or aftermarket vendor. If you have to hire someone to do the welding, finding a replacement may be the cheaper alternative as well.

Be sure to check the integrity of the intake and exhaust manifolds. Small holes in the intake manifold can sometimes be repaired with something like J-B Weld. However, cracks in the exhaust manifold will need to be welded by a capable welder.

Governor Overhaul

While it doesn't come in contact with the fuel, the governor plays an important part in the fuel-delivery system and needs to be inspected as part of your overhaul. In general, the governor uses a rotating mass applied against a spring to adjust the carburetor throttle shaft and thus regulates the engine speed around a set point established by the throttle lever position.

On virtually all Farmall and McCormick-Deering tractors, the governor is a centrifugal flyweight type driven by a gear in the timing gear train. The first step in governor inspection is checking for any signs of malfunction. Symptoms can include the engine idling too fast or not idling down when the throttle lever is moved to the idle position; surging; over-revving; the engine not reaching the specified top speed; engine speed control that is erratic; and delayed reaction or sluggish response to changing load conditions or throttle movement.

Before attempting any disassembly or governor adjustments, inspect all linkages and link rods for free movement and the absence of any bends or binding. If necessary, free up and align all linkages to remove any binding. If additional internal repair or inspection is required, follow the instructions in your repair manual for governor removal, overhaul, and/or adjustment. The basics for governor overhaul include inspecting and replacing any defective bearings, seals, and drive gears. Also, ensure that the flyweights move freely without binding.

If you're not sure of your ability to overhaul or rebuild the governor as instructed in your repair manual, you might consider sending it out to one of the shops listed in the appendix that specializes in governor restoration. Considering the role it plays in controlling engine speed, you want to make sure it's done right.

When reinstalling the governor, make sure the timing marks on the governor drive gear and camshaft gear are in register. Finally, tune the speed adjustment as necessary to limit the engine speed to the rpm rating specified in your service manual.

The governor on most Farmall models is on the shaft that drives the distributor or magneto.

Inspection of the governor should include making sure the flyweights move freely.

Be sure all linkages between the governor and the carburetor are free of binding and operate freely.

During engine repair, it's important to make sure drive gears are correctly timed, including the gear that drives the governor/distributor.

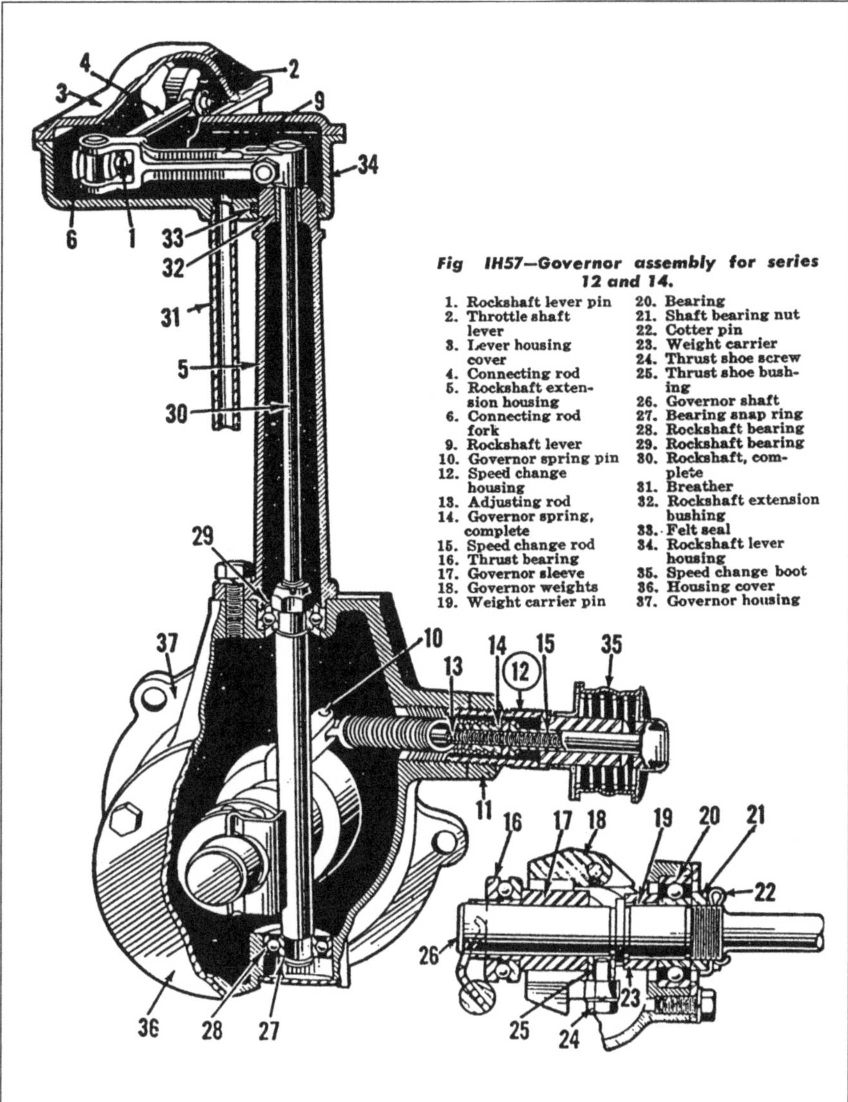

The governor assembly on the F-12 and F-14 was a little more complex, with more linkages, than that of the letter series tractors that followed. (Courtesy of Jensales)

Chapter 14

Cooling System

For those used to today's liquid-cooling system—which normally includes a radiator, pressurized radiator cap, thermostat, water pump, and fan and fan belt (or, on the newest models, a thermostatically controlled electrically driven fan)—the cooling system on most early model tractors appears rather primitive.

Most early Farmall tractors used a simple thermo-siphon circulating system, which consisted of a radiator and two hoses that connected it to the engine block. In fact, the Cub used a thermo-siphon system during its entire production history. Because hot water is lighter, it is forced upward from the engine, through a narrowing passage, to the radiator. Here, it sinks as it is cooled, only to exit at the bottom and return to the engine.

Cooling System Inspection and Repair

You probably had an opportunity to evaluate the cooling system to some extent when you purchased the tractor or during the troubleshooting operation. Perhaps you even had a chance to start the engine and let it run long enough to see if there were any water leaks, problems with overheating, or traces of oil in the coolant.

Unfortunately, radiator cores tend to clog up with rust, lime, or other mineral deposits and the fins plug up with weeds, seeds, and debris. In addition, the metal headers often corrode away after years of use, and the seams become moist with residual antifreeze.

Hence, it's best to start cooling system inspection and restoration at the front of the tractor, at the radiator. The first thing you should do is check the front and rear of the radiator for a buildup of bugs, seeds, weeds, and so on. A strong stream of water sprayed from the back side, or fan side, of the radiator will remove a lot of the debris.

Next, check for moisture around the radiator core and headers. These areas tend to rot out if the tractor has sat dry for a long period of time. If there is leakage, the area will be moist, and perhaps even smell sweet if there is antifreeze in the system. If the leakage is minor, you can sometimes take care of the problem by adding one or two cans of radiator "stop leak" material.

If there is substantial leakage, however, it's best to remove the radiator and have it professionally serviced. Considering that you may have to remove the radiator anyway as part of the tractor or engine restoration process, you may want to think about taking it to a professional, just to have it flushed, flow tested, and checked for integrity. The other option is to simply replace the radiator core as part of the restoration process.

For several years, International Harvester tractors used a simple thermo-siphon circulating system for engine cooling. In fact, the Cub used a thermo-syphon system during its entire production period.

Whether it's removed or left on the tractor, one of the most important steps in cooling system restoration is checking the integrity of the radiator.

Some radiators, especially those used on thermo-siphon cooling systems, can be split into three sections. This allows you to clean or even sandblast the upper and lower cast housings.

This radiator core—which appears to have been patched more than once—is definitely in need of replacement.

This radiator was, fortunately, in reusable condition. However, the overflow tube had broken loose from the filler neck and had to be cleaned and resoldered.

Shutters and Curtains

The cooling system on a large number of early Farmall tractors was designed to do more than just cool the engine. Any all-fuel model, designed to burn both kerosene and gasoline, must have the ability to run hot on demand in order to properly vaporize the less volatile kerosene.

For this same reason, most kerosene tractors include a small gasoline tank that is used for starting the engine. Once the engine and intake manifold are hot enough, the fuel system is switched over to kerosene or fuel oil. That means the operator has to continually keep an eye on engine temperature—not just keeping it cool when necessary, but keeping it hot enough, too. To assist in this endeavor, International Harvester equipped most all-fuel models with a set of shutters that are opened or closed from the operator's seat.

To get the tractor warmed up quickly, the operator simply closed the shutters and deprived the radiator of air circulation. However, there were times, such as when driving back to the farmstead from the field or when moving between fields, that it again became necessary to close the shutters to keep the engine temperature elevated and the engine running smoothly. Since most restored tractors are continuously operated on gasoline, the operation of the shutters is no longer a concern. The main role now is appearance.

Unfortunately, original temperature control devices easily suffer the effects of neglect. Over the years, dirt and chaff would naturally collect between the bottom of the shutters, the radiator, and grille. As a result, the shutters on a lot of restoration-quality tractors have since rusted out as moisture collected in the residue and did its damage. The good news is that, like many other once rare parts, curtains and shutters for most Farmall models have been reproduced and are now available from a number of aftermarket vendors.

In order to raise the engine temperature to a level sufficient for kerosene ignition, all-fuel tractors were generally equipped with a set of shutters or a curtain in front of the radiator.

Radiator Cap

There are basically two types of radiator caps used on Farmall tractors. The earliest type consists of a simple cap that covers the radiator opening and is held in place by a knob-operated mechanism that locks the lid to the housing. Ironically, it's the same cap that covers the holes in the engine block that provide access to the piston rod bearings on the Regular and early F Series models, including the F-20 and F-30. Essentially, its only purpose is to keep some of the steam in and the dirt out. Due to the simplicity, about the only thing you'll need to do is clean the cap, replace the seal, if one was used, and make sure the locking mechanism works correctly.

Before the advent of pressurized cooling systems, the radiator was simply equipped with a cap held in place with a rotating latch. Ironically, the cap that International Harvester used for the radiator was the same as the ones used for the engine block access hole covers.

By the 1940s, International Harvester followed the industry by equipping its tractors with a pressurized cooling system. Like the cooling system found in modern tractors and automobiles, it uses a radiator cap designed to raise the pressure in the cooling system so the coolant boils at a higher temperature. This, in effect, accomplishes two purposes. First, each pound of pressure raises the boiling point by approximately three degrees Fahrenheit, which allows the engine to operate at a higher temperature. Second, since there is now a greater difference between the water temperature and the air temperature, the radiator can operate more efficiently. It is possible, in fact, to find a few tractors, such as late production Model C units, that are equipped with a pressure-type radiator cap, but still have a thermo-siphon cooling system. By the time the Super C came along, though, it was equipped with both a water pump and pressurized radiator cap.

On systems equipped with a pressurized cap, check to make sure the bottom of the cap is clean and fits snugly into the filler neck. Check the rubber bottom for swelling, nicks, or cracks. Also check the filler neck for uniformity on the sealing surfaces. A warp or hairline crack will cause pressure to leak out when in use. Finally, make sure any replacement cap has the proper pressure relief rating. If the relief setting is too high, you run the risk of blowing hoses or the radiator core, especially if the core is weak in the first place.

Fan

The fan and fan drive system on most vintage Farmall tractors are pretty basic by today's standards, as there were no such things as thermostatically controlled electrical fans.

Hence, restoration consists of little more than checking the integrity of the blades to make sure the attachment rivets are tight and the blades haven't been bent, and checking the fan shaft bearings to make sure they're in good condition.

Hoses

On many tractor restorations, the biggest problem with the cooling system—and many times the only problem—is the condition of the hoses. Hoses that are hard, brittle, or cracked need to be replaced. Keep an eye out, too, for small patches of moisture on the hose surface. If discovered, gently knead the area while looking for a hairline crack or pinhole. Such areas tend to leak only when the tractor is at operating temperature and under pressure, making them difficult to locate.

Also, look for hoses that have swelled up because of oil contamination. They feel greasy and spongy when kneaded. Replace any hoses that are marginal. While you're at it, it's a good idea to change the hose clamps, too, since dirt and grit can keep them from being sufficiently tightened to seal water.

As a final note, you'll want to consider how the tractor is to be used before replacing the hoses and clamps with conventional equipment. If you're restoring the tractor as a show model, you may not want to use the type of hose clamps that employ a slotted band and a worm-style tightener. More than likely, the original hoses were held in place by either the old-style spring-loaded wire clamps or wire clamps that tightened with a single screw. They may not work as well, or hold as tight, but if you're going for an authentic look, you need to go all the way.

Always replace any hose sections that are swelled because of oil contamination or feel greasy and spongy when kneaded.

Water Pump

As was discussed earlier, concerning the thermosiphon cooling system, there's a good chance the Farmall or McCormick-Deering tractor you're restoring won't even have a water pump. If your tractor is so equipped, which was the case by the time the H, M, and Super C models came along, the most likely problem you'll encounter is a water leak or worn bearings. If it's a matter of repacking the pump, packing is available in split segment 1/4-inch thick. However, repacking is best done by first removing the pump from the engine, removing the driver pin and drive and unscrewing the packing nut to remove old packing and replace with new.

A leaking seal is generally indicated by coolant leakage at the drain hole in the pump housing. Unless the impeller blades or an internal divider have been completely attacked by rust, you can usually rebuild the unit with new seals and bearings, although the shaft and

Unless the impeller or an internal divider has been destroyed by rust, you can usually rebuild a water pump with new seals and/or bearings.

bearing assembly are available as a preassembled unit only on certain models, including the C and Super C.

If an impeller blade is rusted pretty badly, you may be able to find somebody to rebuild it with a welder once you get it cleaned up. Of course, salvaged water pumps aren't that difficult to find.

The first step in rebuilding the water pump is removing it from the tractor. It's important that you inspect the water pump shaft, assuming it is being reused, to be sure it is smooth and free from rust. Otherwise, it won't be long before you're replacing seals again. If necessary, use a piece of emery cloth to smooth the shaft where it fits against the seal. A small amount of grease on the pump shaft will prevent damage to the water seal as the pump is being reassembled, especially if the impeller has to be pressed back onto the shaft.

Finally, make sure the drain hole in the bottom of the housing is kept free of dirt, grease, and paint so that any water that may leak past the seal can drain away.

The tension on the water pump drive belt can be adjusted by simply turning the front half of the drive sheave clockwise or counterclockwise and locking it in place with the set screw.

Thermostat

Assuming your tractor is new enough to have a water pump, you'll need to make sure the thermostat is operating correctly before you finish the restoration project and put the tractor into use. Obviously, you can tell a lot about its operation by looking into the top of the radiator with the cap removed. (It should go without saying to never open the radiator cap when the engine is hot.)

As soon as the water gets hot enough to open the thermostat, you should see water start flowing into the top of the radiator from the upper radiator hose. As a reference point, most letter series tractors with a non-pressure cooling system have a 130–155-degree thermostat. All other models, including a C and Super C equipped with a water pump attachment, have a 165–190-degree thermostat.

If this is not the case, you have a couple choices. You can test the thermostat by placing it into a pan of water on a stove and watching for the diaphragm to open as the water heats up and attains the temperature at which the thermostat should open. Or, considering the price of a new thermostat and the age of the tractor, you may just want to replace it with a new one. Either way, you'll need to remove the old thermostat. On gasoline and all-fuel models, the thermostat is located in the engine water outlet casting.

One exception to the above procedure, however, is the F-30 and W-30. Both of these models use a thermostat that employ a bellows that opens or closes a thermostat valve in the thermostat cage. In this case, it's necessary to adjust the bellows setting with an adjustment nut and then lock it into position with a jam nut.

Properly tensioned drive belts that are in good condition are essential to trouble-free engine cooling.

Restorers of Farmall Regular models and early F Series models will be pleased to know that reproductions of the old fabric-type fan belts are now available.

Tractor restorers who are striving for originality are always in search of the old-style wire clamps that were first used to clamp hoses in place.

Belts

Most vintage Farmall and McCormick-Deering tractors only have one belt—or at most two—that normally runs the fan/water pump and the generator. However, you need to make sure it is in good condition and not slipping, or the whole cooling system can suffer the consequences.

To check the belt(s), twist it around in several spots so the bottom and one side are clearly visible. Look for signs of cracking; oil soaking; hard, glazed contact surface; splitting; or fraying. Replace any belt showing these symptoms. This may be a little more costly or difficult, though, if you have a Farmall Regular or 10-20. These models used a flat belt made of a canvas-type material. You'll be lucky to find an original in good condition, but fortunately there are reproduction belts available.

Once you've checked the belt condition or replaced it, make sure you adjust it for the proper belt tension. A belt that is too tight can cause premature wear on the bearings, while a belt that is too loose can slip, squeal, or cause other problems.

To preserve the restored cooling system, be sure to add an antifreeze solution containing a rust inhibitor as a final step.

Chapter 15

Sheet Metal

Thanks to the huge interest in antique tractor restoration, today's Farmall enthusiasts have more choices than ever before when it comes to repairing or replacing sheet metal. Due to the growing interest in the restoration hobby, vendors now offer hundreds of sheet metal parts as aftermarket reproductions. Leaf through any issue of the *Red Power* magazine and you'll find ads for everything from complete bare-metal hoods to replacement fenders, dash panels, grilles, PTO shields, etc.

In addition, there are plenty of sheet metal parts available through salvage operations, such as Dennis Polk Equipment or Central Plains Tractor Parts (see appendix listings). Finally, there are a number of businesses, including many Case IH dealerships, that sell what is commonly referred to as new old stock (NOS). These are old parts that have never been used, but instead have been stored in a warehouse or stock room and have only recently been "discovered" and put into circulation.

Paint Removal

Due to the availability of so many new parts, some restorers like to replace any sheet metal parts that they can with new. For others, particularly those on a tight budget, it's more economical to strip and repair existing sheet metal components, unless they are extremely rusted or wrinkled.

When it comes to using a sandblaster to strip sheet metal, you'll find differing opinions. There are some restorers who sandblast all the sheet metal and every bit of the frame prior to a restoration project. And there are others who wouldn't take a sandblast nozzle anywhere near the sheet metal, regardless of how much elbow grease it saved.

Jim Seward, a tractor restorer from Wellman, Iowa, who also manages the body shop for a local General Motors dealership, is one who uses a sandblaster on everything. He insists sandblasting sheet metal is one of the quickest and most efficient methods available for removing paint and rust, providing it is done delicately. In response to people who say, "sandblasting will blow a hole through weak metal," he says, "If the metal is so weak that you're going to blow a hole through it, it needs to be repaired anyway."

For anything but the heaviest gauge steel, make sure you or the commercial operator use fine silica sand or glass beads. Keep plenty of distance between the nozzle and the steel to avoid warping or stretching the part. And always keep the nozzle moving. If the metal gets too hot, it will end up having ripples that are virtually impossible to remove.

Jim Deardorff, who cleans and paints old steel around his home in Chillicothe, Missouri, says he has discovered an even better way to sandblast sheet metal. He has developed "Classic Blast," a special blasting mix made up of aluminum oxide and ground black walnut shells. Using the product in a closed-top sandblast pot, which uses a vacuum to pull the media into the chamber, he says he can reduce the pressure to as little as 35 pounds and still clean fragile parts without damage. To prove it, he often demonstrates the effectiveness of his sandblast method by removing the paint from an aluminum pop can.

Another cleaner that is widely used by a number of restorers, particularly to remove paint from sheet metal parts, is a quality aircraft metal stripper. Just keep in mind that chemical strippers are generally toxic and require adequate ventilation. Also, you need to make sure all traces of the stripper have been removed prior to painting the part or the tractor. Otherwise, the new paint will soon strip off, too.

If you can find someone to do it for you, another good way to strip paint from sheet metal parts is to dip them in a caustic soda bath. A lye solution has the added benefit of removing any grease that may be on a part.

Your local automotive parts dealer should be able to direct you to a wide assortment of paint strippers for cleaning sheet metal down to bare metal. Many restorers say they have the best luck with aircraft remover (far right).

While it's not as easy as using a sandblaster, Walter Bieri of Savannah, Missouri, finds that a wire brush on an electric grinder can take off paint and years of accumulated rust.

Repairing Sheet Metal

Once you have all the paint off, the first thing you're going to notice are all the dents, dings, scratches, and rust spots that need to be filled in, pounded out, and otherwise hidden from view. You may even have to splice in one or two pieces of sheet metal, create a new bracket, or in the worst-case scenario, fabricate a whole new sheet metal section.

Dents and Creases

Let's start with the dents and creases. If a dent or crease is more than 3/16 inch deep, it's best to smooth it out with a body repairman's hammer and dolly. Do not simply fill it in with body putty. Bondo might be fine for automotive repairs, but the vibration that is inherent with tractor operation can cause body putty to pop right out of a deep dent, leaving you with an ugly hole that will require more work and a new coat of paint.

If there aren't a lot of dents to take care of, you might be able to get by with a ball-peen hammer and a mallet or large hammer to back up the piece. The thing you have to keep in mind is that the original metal was stretched as the dent was created. Hence, you may have to shrink it as it is straightened. One way to do that is to heat the spot with an oxygen-acetylene torch before beating out the dent.

Another technique, particularly if you are trying to remove a sharp crease, is to drill a series of small holes (approximately 1/16 inch in diameter) along the crease. This will allow the metal to shift as it goes back into place. The holes can be filled later with epoxy or plastic putty.

Repairing Holes and Rust Areas

In some cases, you may need to cut out an old piece of metal and weld in a new piece. First, remove all the paint from the area to be worked, if you haven't already done so. Next, make a clean cut around the damaged area so you have removed all the bad metal and have left a clean, solid edge. It's important that you cut beyond the damage, because when you take the pieces to a welder, or do the work yourself, any thin, pitted surfaces will self-destruct. You'll also want to remove the section in a shape that will be easy to reproduce. A square section with clean right angles works best.

Now, find a scrap piece of sheet metal that is the same thickness or gauge as the original piece. The biggest mistake people make at this stage is using a slightly thicker or thinner replacement piece. You may also want to trace the hole you have created onto a piece of cardboard and make yourself a template. This will be particularly helpful when you go to cut out the new piece.

If necessary, bend the new piece to match the contour of the hood, grille, or fender where it is being installed. Finally, clamp or tape the new piece into position and tack weld it in place. Then, finish welding around the splice, being careful not to heat the area to the point it warps or disfigures the sheet metal. You'll want to hide as much of the weld as possible, too. Some restorers simply use a series of spot welds around the patch and fill in the seam with body putty a little later. Throughout the process, be careful not to set the welder at too high of a temperature and burn through the sheet metal you're trying to repair.

To complete the patch, grind the welds down to remove any high spots and fill the area with J-B Weld, body filler, or Bondo. This will also fill any rust pits and gaps that have been left. Once the filler has hardened, you can sand it down to the point where the patch is flush with the clean sheet metal, using finer and finer sandpaper to finish the surface.

If you're dealing with rather small holes, you can often get by with putting a piece of fiberglass cloth on the back side of the cavity and filling it with Bondo or body filler. It may take several thin coats before you get the hole filled to surface level. At that point, you can sand and treat it just as you would any other patch.

You can add as many coats of paint as you want; and you still won't hide the dents and wrinkles in the sheet metal on this tractor. In fact, a coat of paint just makes them show up better. It would have been better to replace the panel or spend some time on body work before getting to this point.

There may be times when it is necessary to splice in a new piece of metal to replace a rusted-out area. The answer is to spot-weld a metal piece of comparable thickness in place and then cover the patch with body filler. The entire area can then be sanded to a thin layer that only covers imperfections. You'll notice that was the case on this F-12 hood, as evidenced by the welds on the back side.

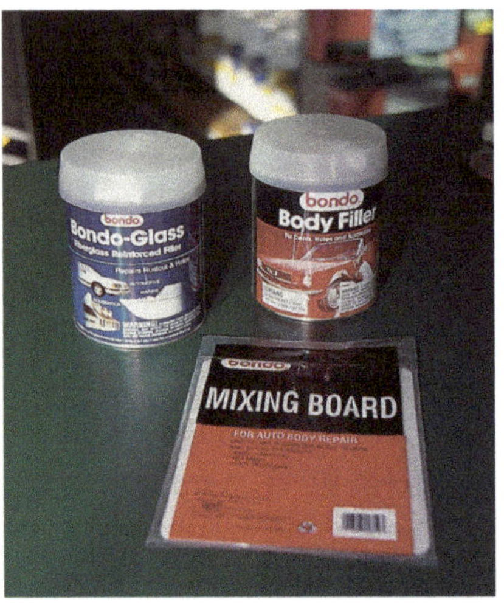

Your local auto parts supplier can direct you to everything you'll need to fill and cover minor imperfections. Just remember that body filler or Bondo should not be used to correct deep dents or creases.

Small dents in sheet metal can often be removed with a ball-peen hammer and a mallet or large hammer to back up the piece. On stubborn dents, though, it may be necessary to apply heat to help shrink the metal back into place.

When mixing body filler, it's important to get the right mix of material and hardener. Otherwise, the material may begin to harden before you have a chance to spread an even coat.

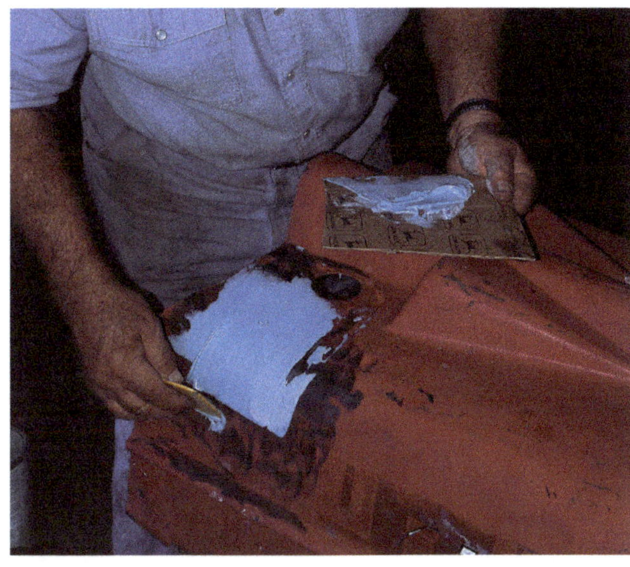

A thin coat of body filler is ideal for filling and covering all the pits left by rust. It should not be used, though, as a substitute for the body work needed to remove dents and creases.

Whether you're sanding by hand or with a power sander, using long sanding boards will help ensure a smooth surface when finishing body filler or a coat of primer.

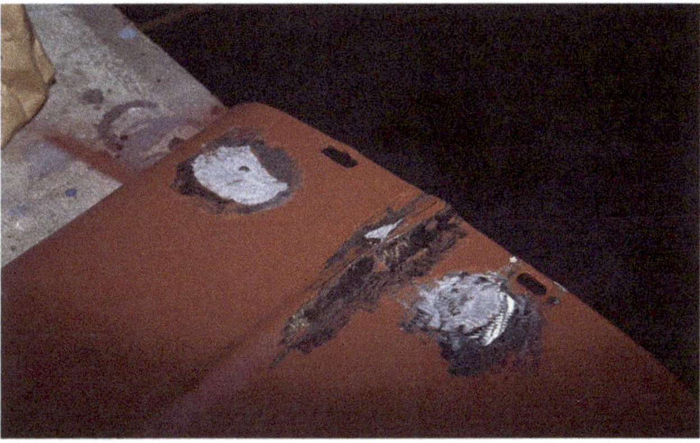

Using welds and some body filler, a couple holes on this hood have already been repaired. The bend in the raised seam still needs a little work, though.

A circular sander may be needed to clean up old paint or rust in some of the tight spots.

The engine, frame, and powertrain on this industrial H have already been stripped of paint via sandblasting and are ready for a coat of primer.

Because of its somewhat fragile nature, the corrugated screen used in many of the Numbered Series tractor grilles is often torn, dented, or gouged. Fortunately, replacement material is available.

The grille may still need a little work, but this hood from an H has already been cleaned and sandblasted in preparation for painting.

The fuel tank on this H has also been stripped and cleaned in preparation for a primer coat.

The instrument panel on this Farmall Letter Series tractor has already been stripped and masked and is ready to paint.

Chapter 16

Paint

Regardless of how good of a job you did on restoration and repairs, the first thing people are going to notice on your restored tractor is the paint job. Small imperfections may not even be visible at this point, but rest assured, once they're covered with a coat of paint, they will show up like neon lights. Consequently, it's important to take your time and do it right. The first step is to lay down a foundation of primer, followed by multiple coats of enamel.

Before you do that, though, it's important that you make sure the surface is clean, especially the sheet metal. Jim Seward, a tractor restorer from Wellman, Iowa, who manages the body shop for a General Motors dealership as a career, recommends you wipe down all sheet metal surfaces with a wax and grease remover solution prior to painting. Once the surface is dry, follow that with a tack rag.

"Try not to touch the metal any more than you have to," he says. "Even the oil on your hands can affect the way the paint adheres."

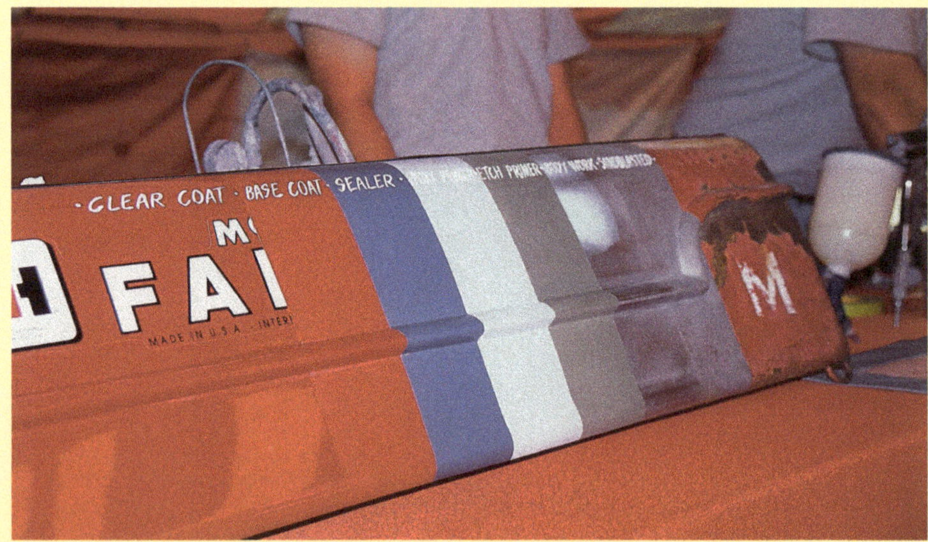

This demonstration panel, owned by Jim Seward from Wellman, Iowa, shows the transition of a hood (right to left) from original finish through bare metal with some filler, etch primer, epoxy primer, sealer, base coat paint, and clear coat.

The Primer Coat

Once the body and frame have been cleaned and prepared, and the sheet metal has been stripped, smoothed, and filled as necessary, it's time to apply a quality coat of primer. Some restorers recommend applying a coat of etch primer immediately after the tractor frame has been cleaned, even if it won't be painted for a while. While the primer isn't guaranteed to prevent rust, it does make it easier to clean away any grease or oil that shows up.

That brings up another tip. Most experienced restorers insist it is best to start up the tractor and drive it around a little, if possible, before applying a coat of paint. You don't have to have the sheet metal or fenders back on yet. The goal is simply to find any leaks now, and to discover unresolved problems in the transmission or engine before that shiny coat of paint has been applied.

The primer stage also gives you the opportunity to take care of a lot of the imperfections that may remain after most of the body work has been completed. By putting on two or three coats of primer and sanding between each coat, you can easily fill a lot of pits and crevices.

Choosing the Right Primer Type

There are a number of different types of primers that can be used to prepare, fill, or seal the surface prior to the final coat of paint. Each has its own unique role and application. For example, while epoxy primers protect components from new rust, urethane primers tend to form a harder finish. However, you need to ensure that the primer you select is not only compatible with any paint that remains on the tractor or engine, but also with the paint you have selected to finish the project.

Hermie Bentrup, a paint specialist with Auto Body Color in St. Joseph, Missouri, explains that the type of primer you start with depends to some extent on the type of finish you're covering—old paint, bare metal, or cast iron.

Epoxy Primers

If you're shooting primer over bare metal or cast iron that could be exposed to the weather before you get it covered with additional primer, Bentrup generally recommends an epoxy or self-etch-type primer. This is particularly the case with parts of the frame or cast wheels that have been sandblasted and are in no need of further work or sanding.

You'll find several different kinds of primers on the market, including filler primers and those that are self etching. Your paint supplier can help you select the best type for your needs and application.

This clutch housing cover has already been hit with a primer to prevent any rust before it's time for reassembly.

Epoxy is the easiest to use because it combines the qualities of a metal etch, a primer surfacer, and a primer sealer in one product. A self-etching primer, on the other hand, is basically a phosphoric-acid-type etch.

Self-etching primers have to have another primer over the top of them, Bentrup adds. It can't be an epoxy primer, but it can be a urethane primer. You cannot paint directly onto a self-etching primer, because the paint won't adhere. That's why painters generally recommend that tractor restorers go over bare finishes with an epoxy primer, since it will give you a more durable finish, and it will etch aluminum and metal in one shot.

Epoxy primer can be used in one of two ways. It can be used as a primer-sealer, where you spray it on, wait 15 to 20 minutes, and start top coating with your color. Or it can be used as a primer-surfacer to cover minor flaws in the surface. In this case, you will want to put down two to three coats, giving it 15 to 20 minutes between coats. Then, wait at least 6 hours before sanding the surface.

Finally, epoxy primer can be sprayed over the top of old lacquer paint—which was often used on older tractors—without a problem. It is a good idea, however, to seal the original lacquer just to be sure the two surfaces remain compatible. It is not recommended that you ever put lacquer on top of enamel.

Urethane Primers

Although urethane primers are very popular with restorers due to their hard finish, they do not include any kind of chemical agent to prevent rusting. Therefore, you either need to ensure that the surface is completely free of rust before you apply a urethane primer, or you need to lay down a coat of epoxy primer or an etching primer and put the urethane over it. Otherwise, you may find rust popping through the surface four or five months down the road.

Since urethane is only a primer-surfacer, you'll also need to apply a sealer of some kind before you paint. Again, a coat of epoxy primer over the top will serve the purpose. Plus, a coat of epoxy will serve to bridge any scratches in the urethane that have been left after sanding and leave a smooth surface for the top coat color.

The only thing you shouldn't do with urethane is put it under a top color coat of lacquer.

Filler Primers

For even more complete coverage of imperfections, you might want to follow the lead of other tractor restorers and use a filler primer, or heavy-bodied primer, which fills in pits even quicker. You may even want to use a little spot putty to eliminate any air pockets and holes found in the filler primer after it dries. Between coats of primer, be sure to use a fine-grit sandpaper, such as a #240 grit, to smooth the surface. On the last sanding before painting, you'll want to use an even finer sheet, like #400 grit, if you're using enamel and up to #600 for urethane or lacquer.

Some show-tractor enthusiasts have been known to apply up to fifteen coats of filler primer on a sheet metal part that has had a lot of work, just to get a glass-smooth surface. Between every three coats, they use long sanding boards to smooth the finish. This helps prevent high and low spots from being formed by the sander itself. Another trick used by some body specialists is to switch between different colors of compatible primers so they can better identify hills and valleys in the finish. Be sure to finish with finer and finer grades of sandpaper, ultimately ending up with wet sanding paper.

Although many so-called filler primers are actually lacquer-based primers, some urethane primers are also called filler primers. One of the beneficial characteristics of a urethane filler primer is the fact that when you sand it, the material will actually reflow and close up. Urethane primers are usually 1 to 2 millimeters thick per pass, says Bentrup. So if you go around the tractor three times, you've got 6 millimeters of primer on there, whereas most new cars have about 7 millimeters total, including all primers and paints. In the end, you'll have a thicker, tougher finish that will last for years.

It's important to note, however, that any urethane primer will need a sealer coat before the top coat of color is applied.

Sealing Primers

The final step before applying a coat of paint should involve applying a coat of sealing primer. This closes the surface and prepares it to accept a coat of paint. Before you apply the sealer, though, use a tack cloth a second time to get the surface extra clean. You don't want to seal in any dirt or sanding dust.

The Color Coat

If you think all Farmall tractors were red, the first thing you need to do is brush up on your International Harvester history. Like most other tractor brands, IH started painting its first tractors a rather drab color. In the case of IH, it was battleship gray, or Tractor Gray, as the company called it. Other manufacturers painted their models dark green, blue, and their own shades of gray.

Allis-Chalmers was one of the first to make a change when the company switched from green paint to its infamous Persian Orange in 1929—supposedly after Harry Merritt, manager of the tractor division at the time, saw a field of brilliant orange poppies while on a trip to California. Of course, that's just one story concerning the origin of the paint color. By the 1930s, other companies were changing to brighter paint colors, too. International Harvester made the change in November 1936, when the company introduced Farmall Red on its 1937 models.

Of course, Farmall Red wasn't the only color used on Farmall tractors over the years. Every once in a while, you'll run across a tractor painted white. The most well-known and documented was a line of demonstrator tractors that were painted white for the Mid-Century Sales promotion in the early 1950s. Most of them were Cubs, Super As, and Cs; however, rumors persist that there were also a few H and M models painted white. At any rate, the demonstrator tractors were also adorned with stick-on stars that pointed out sales features, as well as cardboard wheel inserts and advertising placards that highlighted Farmall benefits.

While any true white demonstrator tractor is surely a collectors item, it's difficult to tell anymore which ones are really authentic. While a few models have been painted white to imitate a demonstrator tractor, others that were the real thing were repainted red after the promotion was over and sold to farmers through local dealerships.

Through the years, Farmall and McCormick tractors also featured different trim colors, including cream, blue, and black. Wheels, meanwhile, were either red or galvanized; although most restorers reproduce the galvanized look with silver paint.

International Harvester also complicated the issue with a program it sponsored in the 1950s, in which customers could have the local dealer repaint their tractors. In many of these cases, employees at the dealership weren't as careful about masking off certain parts as the factory workers. On Letter Series tractors, for example, engineering specifications called for leaving electronic components unpainted. This included the magneto, spark plugs, spark plug and coil wires and, in most cases, the distributor cover. However, it's not uncommon to see many of these parts painted in red today—especially the distributor cap and some wires.

If you really want to do your research on the correct paint scheme, you can always go to the Wisconsin Historical Society website and search through the records. All of the IH paint committee decisions from 1924 through 1960 are listed here. For example, if you have a 1931 F-20, you'll discover in a paint committee memo date June 5, 1930, that the drawbar on Farmall tractors was to be painted Harvest Red instead of Tractor Gray, as they originally had been. The decision was made at the request of the works manager who stated that "because of the physical layout of the Farmall equipment," some money could be saved by painting the drawbar the same color as the wheels. The trunion castings and braces were to remain gray, as in the past. However, if you look around at many tractor shows, most Farmall tractors continue to have a gray drawbar, regardless of the year.

Another memo dated September 24, 1946, states that all tractor weights are to be painted the same color as the tractor. No notation is made as to what color they were previously.

Until November 1936, when the International Harvester introduced Farmall Red on its 1937 models, Farmall tractors were painted battleship gray, known in the factory as Tractor Gray.

Industrial standard tread and utility tractors were often painted yellow, just as they were with a number of tractor brands.

Selecting the Right Paint Type

Just as you did with the primer, you'll need to decide what kind of paint you want to use on the tractor. While some restorers prefer acrylic enamel, others opt for the new polyurethane finishes. Although lacquer is seldom used anymore, you'll find it covered here, too, primarily as an answer to questions that often come up among restorers.

According to Jim Seward, there are basically three types of paint in use today. As he relates, they include "Implement" or synthetic enamel, acrylic enamel, and base coat/clear coat. Although lacquer was often used in the past, Seward, Bentrup, and others say it is a thing of the past for reasons explained.

Lacquer Paints

One of the things that made lacquer the choice of previous generations is the reason you seldom see it used anymore—that is its vaporization and evaporation characteristics. Lacquers tend to dry very quickly because of the rapid evaporation of the solvent used as a carrier. For that reason, the EPA would prefer nobody used them anymore. And, to be honest, few people do.

According to paint specialist Hermie Bentrup, lacquer has other disadvantages, too, including the fact that it dries to a dull finish and must be buffed to bring out a shine. Lacquers are also the most photochemically reactive, which means they fade over time when exposed to sunlight—though newer acrylic lacquers offer improved ultraviolet radiation protection. Finally, lacquers do not withstand exposure to fuel spills and chemicals, as well as other types of finishes.

On the other hand, lacquers are probably one of the easiest of the paint types to apply, Bentrup says. If you make a mistake, you just sand it down and paint right back over it. With enamels and urethanes, there are steps you have to take to recoat it.

Implement, or Synthetic Enamel Paints

Jim Seward describes implement enamel paints as anything that comes in a can from the equipment dealership or farm store. This includes aerosol cans of red, black or cream-colored paint used to touch up parts. That doesn't mean these paints are a poor choice, though. There are literally hundreds of Farmall tractors in the country that have been restored with Number 2150 red paint purchased through the Case IH dealership.

"The one good thing about these is the price," he says. "They're a lot less expensive than either the acrylic enamel or a base coat/clear coat. Unfortunately, they don't have near as much gloss and the chemical bond often isn't as good."

If you can verify that it's authentic, a demonstrator tractor, like this Model A that was painted white for the Mid-Century Sales promotion in the early 1950s, can be a real find.

Acrylic Enamel Paints

Perhaps the most popular paint type these days is acrylic enamel, since it is available in a broad range of colors, allowing it to be custom mixed to match virtually any tractor color. In addition, enamel is relatively forgiving and requires minimal surface preparation. The downside is it takes a little longer to dry and must be applied in multiple, light coats to keep it from running. It also costs about twice as much as implement enamel.

Another characteristic of acrylic enamel, Bentrup explains, is that it dries from the outside in. This means that the underneath side of the coat is still porous for quite some time. As a result, if you spray back over it too soon, the new coat will work its way under the top layer and cause it to lift or wrinkle. Remember, unlike lacquer, which air dries, enamel paints dry and adhere through a chemical process.

For that reason, both he and Seward recommend the use of a hardener, which causes the coat to dry from the inside out and allows recoats without a lift problem—the next coat can be applied as soon as the first coat is dry to the touch. In addition, an acrylic hardener will increase the gloss and provide a more durable finish. Adding a hardener has the negative affect, though, of reducing the pot life of the mixed paint. In the case of enamel, the pot life goes from about 24 hours without hardener to about four hours once the hardener has been added.

Acrylic enamels that match IH colors are available from several different sources. However, the most commonly quoted numbers are from DuPont—those being 7410 for IHC red and 27625 for charcoal gray. Both numbers are in Centari and Dulux formulations. However, Dulux contains lead and will most likely be

When painting a tractor, it's important that you use a thinner or reducer to obtain the correct consistency. Several thin coats are better than one or two thick ones. To improve drying time and strengthen the paint coat, adding a hardener to the paint is a good idea.

fazed out, if it hasn't already. Using the DuPont number, paint suppliers can match other brands to your needs, as well. Other numbers that have been offered by restorers and web site discussion boards include Martin-Senour 99L-8121 (charcoal blue) and 99L-8711 (blue); PPG 32565 (gray); Tisco TP110 (red) and TP900 (cream).

Urethane Paints

Urethane paints, which are actually part of the enamel family, are becoming popular. Among the reasons are the fast drying time compared to acrylic enamel, and the durability and luster that accompanies the hard finish. On the other hand, urethane paints are not available in nearly as many colors as acrylic enamels.

Bentrup notes that there are basically two types of urethane in use. One is a base coat with a clear-coat finish, which is what most of the automotive manufacturers have gone to. In essence, it's a cheaper route, even though the car companies would have you believe it is a superior finish. In effect, less pigment is used to lay down a color and the gloss comes from the clear coats that go over it. As a result, base-coat finishes require at least two to three coats of clear coat for both protection and shine.

The other type of urethane is a single-stage urethane. Like acrylic enamels, there are some single-stage urethanes that don't need to be clear coated, although both single-stage urethanes and acrylic enamels can be clear coated for additional shine and protection. In the absence of a clear coat, two or three coats of single-stage urethane are recommended.

In Seward's opinion, a base coat followed by a clear coat provides the best gloss of any paint coat. It's also one of the most expensive, which means it's often reserved for museum quality restorations. A base coat/ clear coat finish is also very susceptible to scratches, he insists.

"You wouldn't want to just take a rag to a tractor that's been collecting dust at a tractor show," he relates. "You'd see hundreds of tiny scratches in the clear coat the next time you looked across the hood in the sunlight."

A better alternative, he believes, is to wet sand a coat of acrylic enamel with about #1500 grit paper and buff it with polishing compound.

"You can get nearly the same amount of gloss," he says. "It's just a lot more work and you have all the edges to contend with."

Ensuring Paint Compatibility

Although restorers all have their favorite choices of paint type, it's important to ensure that the primer, paint, thinner, and clear-coat protectant are all compatible with each other.

One tractor restorer experienced the frustration of having to repaint a hood and grille three times because the paint bubbled as it dried. After several attempts to figure out the problem, he finally traced it to the hardener, which was either bad or incompatible with the paint he was using.

"We recommend that you stay within the brand that you are shooting to ensure compatibility," says Bentrup. "Just like everything else, there are off-brands that will work. But we try not to recommend them, because if there is a problem, then you've got everybody pointing fingers. If you're within the brand, you know everything has been tested in the lab for compatibility and, unless you've made a mistake, the company has to stand behind it."

As a body shop specialist, Seward agrees, noting it's just not worth the risk in terms of time and paint cost to try and save a little money on the hardener or sealer.

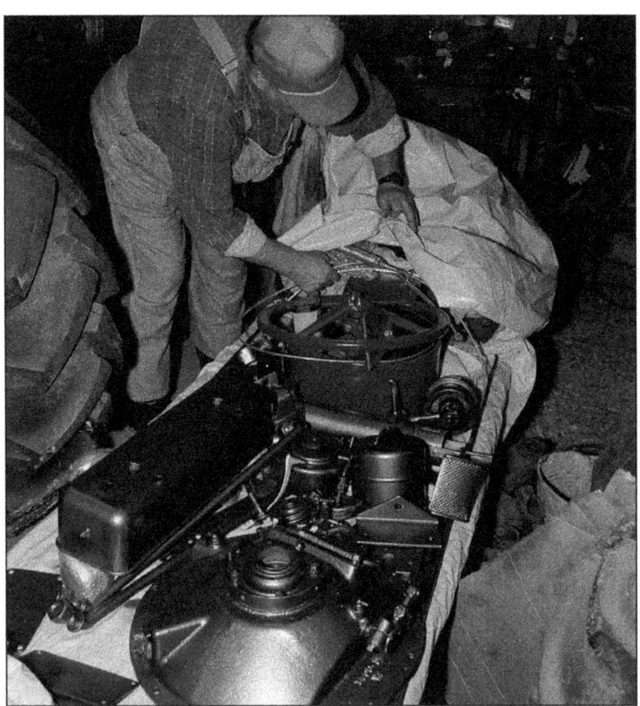

Painting a tractor is much easier if you have the room or time to spread out the parts and paint them individually before trying to assemble them into a finished project—even if you have to cover them until later, as was the case with these parts for a rare Farmall Regular.

Painting Equipment

Selecting the right paint for your tractor is one thing. Gathering the right equipment to protect yourself while completing the task is quite another. In the long run, the latter is really the most important At the least, protective equipment should include rubber gloves, protective clothing, and an approved respirator or mask.

To be safe, use a respirator approved for organic mists, which is the type labeled for use with pesticides. While a charcoal-filter mask may be sufficient for enamel paint, urethane coatings and acrylic enamels to which a hardener has been added contain chemicals known as isocyanates, which are especially toxic. Hence, the use of urethane requires a fresh-air mask and painting suit, or a charcoal mask and a fully ventilated environment.

One of the requirements for a good paint job is a quality paint gun and a compressor with adequate capacity. One of the newest types of applicators is the HVLP, or high-volume low-pressure gun, which helps conserve paint and reduce vaporization.

Remember, too, that paint, thinners, and solvents are highly flammable, particularly when atomized by an air-powered spray gun. So be sure the area is free of any open sources of ignition and keep a fire extinguisher nearby.

Finally, make sure your air compressor and paint gun are adequate for the job. Most restorers use a siphon-type gun that siphons the paint out of a canister or cup and draws it into the air stream. If you want to spend about twice the price for a new gun, you can move up to an HVLP unit, which stands for high-volume, low-pressure painting. This type of gun feeds the paint directly into the air stream, which tends to save paint and generate fewer vapors.

A respirator approved for use with organic mists is one of the most important pieces of equipment when painting a tractor. Remember, too, that paint fumes are not only hazardous to your health, but they are flammable.

Regardless of what type of paint gun you use, make sure your compressor will provide enough air capacity and that you have enough hose to move freely around the tractor.

As a final note, Jim Seward says he likes to make a "paint booth" around the project with drop cloth or plastic sheeting. The plastic not only keeps paint from drifting away from the area, but the plastic holds enough static electricity to attract dust and spray that could negatively affect the paint coat.

Applying the Paint

If you're like most restorers, you can't wait to get the tractor assembled and get it painted. But try to restrain yourself just a little longer. You'll achieve the best results and have the easiest time painting your tractor if it is still disassembled. That means you should look at painting individual sections of sheet metal, as well as the frame and engine, separately, whenever possible. Some restorers even prefer to paint the wheels with the tires removed, rather than masking them, to avoid the potential for overspray on the rubber. Just make sure the paint has had plenty of time to cure before remounting the tires and be ready to touch up any blemishes.

By leaving as many parts off the tractor as possible, you also have the opportunity to paint both sides of a piece in one session. Components like the seat, grille, battery box, and so one can be hung on wire hooks, for example, and coated on all surfaces, without having to let one side dry first. If you're painting a two-tone tractor, like a 350 or 450 that have white panels on the hood, most restorers suggest masking the tractor and painting the white areas first.

It's important to adjust both the spray mixture and the spray pattern. One, of course, can affect the other. In general, several thin coats of paint are better than one or two thick ones. On the other hand, if you get the paint too thin, it can have a dusty appearance that reduces gloss and shine. To attain the right consistency, you'll need to add thinner or reducer. According to Bentrup, these two components do basically the same thing: They improve the spray pattern and the paint's ability to evenly coat the surface. It's just that they're usually referred to as thinners when used with lacquers and reducers when used with enamels, including urethane products. Seward adds that reducers also come in slow, medium and fast speeds. For faster drying times, use a faster reducer.

Once you have attained a mixture that sprays smoothly and evenly, adjust the nozzle to spray an oval that is approximately 3 inches by 6 inches at about 1 foot distance. When painting large areas, spray around the edges first and finish up by filling in the center. Concentrate on moving the gun in a back-and-forth motion to produce a smooth, even coat.

As with many things in life, practice makes perfect. So, if you don't have much experience with painting, start by practicing on a few scrap pieces of metal.

Most restorers and paint suppliers recommend at least two to three coats of paint; although some use up to five or six on sheet metal for extra durability and shine. One tractor restorer, who specializes in museum quality restorations, applies at least seven coats of enamel on sheet metal, putting one coat on right after the other. After he finishes the last coat, he lets it cure for three or four days, then wet sands all sheet metal with #1200 grit sandpaper, staying clear of any edges. Finally, he applies the decals and sprays clear coat over both the paint and the decals to produce a gleaming shine. Any cast parts, however, receive only paint to avoid a glossy, unnatural appearance.

Bentrup notes that one way to ensure adequate paint coverage is to stick a special black-and-white check panel on a masked area before you start painting. Once the black-and-white grids on the check panel are completely covered and hidden, you know you have sufficient coverage; any additional coats are simply building up the finish.

Finally, make sure the timing between coats is sufficient, particularly if you're using an enamel without a hardener (see the explanation about enamels).

On the other hand, since the coats rely on a chemical bond for adhesion, you don't want to wait too long between coats either. Most paint coatings, or your paint supplier, will provide some kind of guidance, so follow the recommendations.

Some restorers like to paint parts of the engine as they're added and sealed to avoid missed or hidden spots later.

Hanging small parts on coat hangers or pieces of wire allows you to paint all sides of the part in one pass.

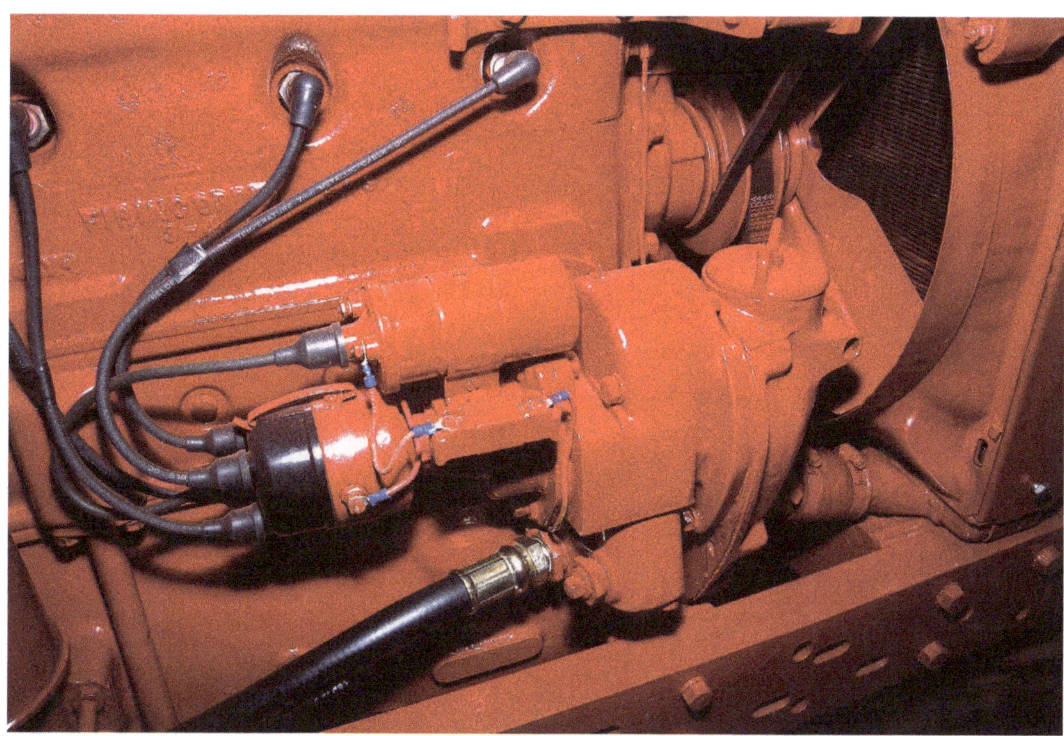

The distributor cover and spark plug wires on this model are correct according to original factory specifications that called for these parts to remain unpainted.

Despite your best efforts, there will be spots that need to be painted by hand or touched up after assembly.

The frame and engine are best painted separately. The same goes for sheet metal. The tractor can then be reassembled once all components have dried and cured.

Wheels on later model Farmall tractors were originally galvanized, but can be nicely restored with silver or aluminum paint.

With all but the hood back in place, this Model 450 is ready for decals.

Chapter 17

Decals, Name Plates, and Serial Number Plates

When it comes to decals, the good news is they have never been easier to apply or easier to find. The interest in antique tractor restoration—coupled with advances in computer-graphics technology—has seen to that.

The reproduction decal business started when a few collectors commissioned a printer to produce a set of decals that they couldn't locate. Soon they found there was a demand for the product and began selling the decals to other collectors. Using modern scanning technology, decal manufacturers are now able to produce decals from drawings, literature, operating manuals, or even pencil rubbings.

The other thing that has changed is the type of decals available. At one time, almost all of the decals on the market were the old water-transfer type, similar to the ones we used to apply on plastic airplane models. After soaking them in water, the backing paper was slipped off and the decal was applied and left to dry.

When International Harvester first started applying decals, they, too, were the water-transfer type. Unfortunately, water-transfer decals have a limited shelf life, since they tend to crack easily. They also begin to weather and crack very quickly after they have been applied to the tractor. As a result, water-type decals are seldom, if ever, used today, especially since technology has helped decal suppliers develop better alternatives.

These days, many of the decals sold and used are made by screen printing onto Mylar plastic. With this type of decal, you simply remove the backing paper, which protects the adhesive, carefully press the decal into position, and remove the front layer of paper that protects the letter surface, much like you would apply a bumper sticker.

Equally popular are vinyl-cut decals. Unlike the Mylar decals, however, vinyl-cut decals are sandwiched between two layers of protective paper. Since each letter or number is individually cut out of vinyl, the paper on the back protects the adhesive, while the paper on the front holds each of the letters in place as they're being applied. Unfortunately, vinyl-cut decals are often a one-shot deal, without any chance for adjustment, since the letters are all separated from each other.

Researching Decal Originality and Placement

Before you get started with any type of decal, it's important to have the right tools and the right information. The latter, however, may be the most difficult to obtain. You may think you know where all the decals go, but even experienced restorers have disagreed over what decals are truly correct on Farmall tractors. Decal configurations occasionally changed from one year to another, even within the same model.

To make matters worse, International Harvester sponsored a tractor repainting program in the 1950s. Whether it was to help dealerships make some extra money during the slow months or to improve the economy when tractor sales were slow, makes little difference today. The program resulted in hundreds of tractors being repainted—as many as forty from a single dealership. With few exceptions, however, the dealership that did the painting installed new decals that were current at the time of the paint job. That means that a 1942 Farmall that was repainted in 1951 received the 1951 decals. The question you have to answer is, Do you want it to look like an original 1942 model, or do you want it to look like the model you acquired?

There are plenty of cases, too, where dealers made modifications to correct a factory problem. If it involved sheet metal replacement, as was the case with early M models to fix a vibration, new decals were provided with the "fix." Again, these were generally current decals, which changed the tractor from its original appearance.

It's important to know which decals are correct before you start this final stage. For example, before the Torque Amplifier became standard, models so equipped had a "TA" with an accompanying "Torque Amplifier" flag added to the model designation decal.

The letter decal on Letter Series tractors is always positioned within the curve on the hood bead. The gap between the decal and the curved bead should be about 1/2 inch and should be equal all the way around the curve.

As a result of all the changes and modifications over the years, John Hunter with Maple-Hunter Decals says he doesn't even try to tell restorers what decals go on which tractors or where. Instead, he lets the customer decide what they want. "There are just too many opinions," he says. "I worked for International Harvester for a number of years and I can tell you the company made hundreds of changes over the years. So, I'm not sure there is anyone who can tell you exactly what is correct."

Maple says he has also seen numerous variations on the way the IH decal and McCormick-Deering or McCormick decals have been used together. Some restorers use an IH decal off to the left in combination with the McCormick name while others use the McCormick-Deering name centered over the "Farmall" decal—without the IH decal.

One option, if you don't have photos of the tractor before it was stripped, is to go to the Wisconsin Historical Society website and review the hundreds of original photographs and color posters that depict Farmall and McCormick-Deering tractors. If you don't find a photo that matches your tractor, you might want to write or e-mail the curator of the McCormick-IHC Collection and ask for a photocopy of any material that illustrate the model and year of your tractor.

One final warning from experienced Farmall restorers is to be careful when buying a package of decals that includes logos and lettering for more than one tractor model. If there are extra decals in the package, people often feel like they need to use all of them.

Tools and Supplies

In one respect, applying decals is no different than overhauling the engine—you need the right tools and supplies before you start. These should include a roll of paper towels; a clean, soft cotton towel; a roll of masking tape or drafting tape; and a rubber or plastic squeegee (you can find these in most craft, automotive, and wallpaper supply stores). A sponge may be helpful, as well. Plus, you might want to have a pair of tweezers handy for holding the edge of smaller decals. If you're using Mylar decals, you'll also need a water tray for wetting the decals before they are applied.

Surface Preparation

The surface needs to be thoroughly dry before applying decals. If you are applying Mylar decals to any painted surfaces, you also need to be sure the paint has cured. Depending on the climate in which you live, this could be anywhere from a week to a month after the tractor has been painted. If a hardener was used in the paint, you may need to wait even longer to make sure the paint isn't going to give off gas bubbles under the decal. Unlike vinyl decals, which have the ability to "breathe," Mylar is impermeable to air and gas bubbles. So any bubbles that form under the decal after application will stay there. The paint surface must be smooth and absolutely clean, as well. If there are any pits or surface imperfections, the decal may not adhere properly. Finally, make sure your hands are clean.

You need to make sure the room temperature is within a comfortable range, too. Decals don't do well when the air temperature or metal is too cold. If you're applying vinyl decals on a hot surface or under a bright sun, they can get too warm and stretch as you're trying to stick them in place and smooth them out. The shop should be between around 60 and 90 degrees Fahrenheit to ensure that the decals adhere correctly.

International Harvester used a variety of name decals on utility, orchard, and wheatland tractors. Some used a "McCormick" decal, while others listed "McCormick-Deering" or "International Harvester," leaving many restorers wondering which is correct. As a general rule, prior to 1949, the "International" decal was used only on industrial tractors. Meanwhile, the "McCormick-Deering" name was used until 1949 when it was simplified to "McCormick" only. The use of the "IH" logo with the "McCormick-Deering" name is more difficult to trace. It's believed to have taken place in 1947 when other numbering systems were changed.

Decal Application

Now that everything is ready, the first step is to hold the decal in the proper location and mark the edges with a piece of tape. You should be able to see the actual decal outline, even if it does have a protective film on each side. A few pieces of tape to mark the bottom edge and a piece of tape on one or both ends will give you reference points.

At this point, it's simply a matter of peeling off the backing and applying the numbers or lettering to the tractor. Still, there are a few tips that will make the job easier, depending on which type of decal you are applying.

Mylar Decals

If you're using Mylar decals, you might do like many veteran restorers and make sure your hands and tools are clean and wet. This will help keep the decal from sticking to surfaces it's not supposed to.

Some restorers, like Iowa's Jim Seward, like to also use a spray bottle filled with water and a single drop of soap to spray the metal surface. In the meantime, fill a cake pan or similar-sized container with water and another drop of soap. Finally, run the decal through the pan of water before placing it on the metal. "The only purpose of the soap is to break the surface tension of the water, so you don't need much," Seward says. "It gives the water a sheeting action, instead of beading up."

Another option used by some restorers, including John Hunter, is to use Windex for the same purpose. Just be sure you use a formula that does not contain ammonia, as it can damage the paint and cause it to fade in sunlight. Also, use the form that comes in a pump-spray bottle. An aerosol can, Hunter warns, produces too many bubbles.

While it may seem that the water or liquid is used so you can remove the decal or shift it around if you make a mistake. "And you can do that," Seward continues. "The main reason you should wet a Mylar decal, is so you can more easily squeeze out all the air bubbles."

Remember, Mylar doesn't breathe, so if there are any air bubbles under the decal once it dries, they're real tough to get out.

Once the decal is in the exact location you want it, use the squeegee to press the decal into place and remove any water and air bubbles from beneath it. Start in the center and work outward. Then, use a soft cloth to dry the surface and remove any adhesive left on the surface.

Vinyl Decals

Unlike Mylar decals, vinyl decals can be applied almost immediately after painting, since the material will allow air and solvents to pass through it. Keep in mind, though, that some decal kits will contain both vinyl and Mylar decals. So you will have to treat the latter accordingly.

"Some decals are only available in Mylar and some come in a choice of vinyl and Mylar," says Hunter. "We let the customer decide which type they want."

When it comes to vinyl decals, some restorers simply mark the position with a few pieces of tape, peel off the backing, and stick the decals in position using the

Don't forget about the small instructional decals that were a vital part of the original model. However, it's important to know which ones were used and where, since decal kits often apply to several tractor models and include extra decals.

tape as a guide. There's another method, however, that can save time and reduce the margin of error. To begin, place the decal in the correct position and "tack" it in place with a few pieces of tape. Once you have ensured it is in the right spot, run a piece of tape across the full length of the top. This piece of tape acts as a hinge for the decal. Next, remove the pieces of tape that acted as a temporary tack, leaving only the top hinge. Now, all you have to do is lift up the decal, pull off the backing and drop it back down into position. If you're working with a long decal, such as a decal that goes along the side of the hood, you can generally cut it into two pieces, separating it between letters, so it's easier to work with. To finish it off, simply press the decal in place, remove the top protective paper and smooth everything with a soft, dry cloth.

Seward says another option, instead of cutting the decal, is to securely tape a long decal in the middle and pull off the paper toward the ends. This allows you to handle long decals one-half at a time.

Don't worry if you have a few little bubbles this time. All you have to do is set the tractor out in the sun. The pores in the vinyl will open up and allow the air to permeate through the decal, leaving a smooth surface. As stated earlier, just don't try to apply the vinyl decals in the sun. They may stretch.

Clear Coat or Not?

Although some restorers like to finish off the decals with a shot of clear coat, others say they never put paint or clear coat over any kind of decal. For one thing, you have to know that the decal can take it and that the protectant won't cause it to lift off the surface. Water transfer decals, for one, can't take it.

Some decals do have a tendency to yellow when covered with clear coat, even if the surface is better protected. Body shop owner B. J. Rosmolen from St. Joseph, Missouri, says he never puts clear coat over any kind of decal or appliqué, including pin stripes on an automobile, simply because it's a lot easier to replace a decal in the future than it is to restore the paint finish.

If you happen to have problems with a decal, such as a letter getting gouged, you can replace a single letter. But if you covered the decal with clear coat, that gloss is not going to be on the new letter. Plus, you're going to have a more difficult time getting it off.

The companies that supply the vinyl for decals don't always want to warranty the product if it's sprayed with clear coat, either. Still, you can find plenty of restorers who have applied clear coat over decals without any problems. In the end, it appears the choice is up to you as the restorer and your willingness to take chances in the interest of appearance.

The first step in decal replacement is determining the correct location, which is often done by measuring the distance from a certain seam or sheet metal edge.

Some restorers suggest temporarily tacking the decal in place and then running a piece of tape across the top of the decal to act as a hinge. If you have a long decal, such as one that reads "McCormick-Deering" or "International Harvester," you may want to cut it into two pieces to make it more manageable.

By lifting the decal on its tape hinge, you can then peel off the paper backing.

Drop the decal directly into position and press it down firmly.

Once the decal has been pressed in place, carefully peel off any protective backing paper. Most Mylar decals, such as this one, don't have a backing paper.

Make sure all the air bubbles have been pushed to the edge and eliminated.

As a final step, wipe the surface with a soft, clean cloth to dry the surface.

Remove all hinge tape or reference tape pieces, and you're finished.

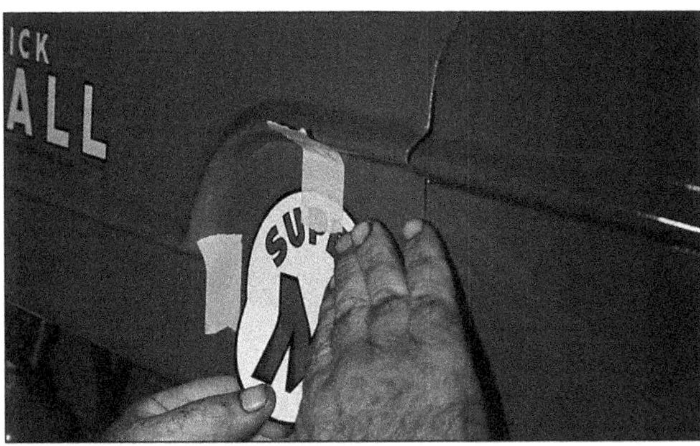

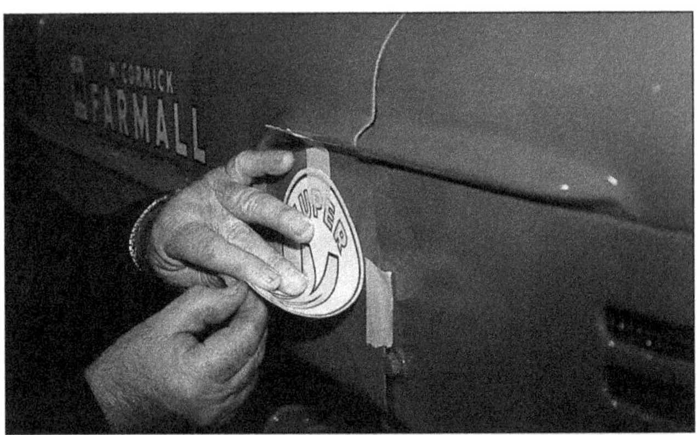

On small pieces, especially those that are round, like the model designation decal, it's helpful to make a hinge at the top, as well as a reference point on each side. This will keep you from rotating the decal, or shifting it right or left, as you drop it in place.

Emblems and Name Plates

As was the case with most all early tractors, a decal was the extent of any adornment or identification on early International Harvester models. However, just as automakers started adding chrome and special emblems, so did tractor manufacturers. In the case of International Harvester, it came about in 1939 with the introduction of the Letter Series tractors.

When Raymond Loewy redesigned the Farmall tractors, his assignment included not only the tractors and their controls, but the company logo, as well. His new design for a logo was based on a capital "H" with a small, dotted "I" in the center. It not only included both letters for International Harvester, but viewed straight on, it was designed to suggest the appearance of a row-crop tractor and driver. On a number of models, the new logo appeared as an emblem on the front of the hood. On other models, or later versions of the same model, the word "Farmall" was spelled out in chrome letters on a distinctive hood emblem.

Unfortunately, both emblems were in a position where they could be easily damaged while working with a loader, or when the owner bumped into something in the back of the machine shed. The good news is reproduction emblems are readily available, which means it's easier to find a new emblem than to restore the original. It will look a lot better in the end, too.

International Harvester didn't limit its use of emblems and lettering to the grille, however. With the advent of the Numbered Series in 1954, virtually all numbers and letters were in the form of raised stainless steel emblems. Whether they appear on a solid red background, as is the case with the 100, 200, and 300, or on a white panel, like that of the 350 and 450, there's nothing as striking as a glossy paint job highlighted with bright, shiny lettering. Again, side emblems are readily available as reproductions and range from around $35 to $50 per emblem.

The best paint and decal job in the world will lack something if the emblem that adorns the grille on many tractors is not restored or replaced.

The "McCormick-Deering" nameplate should generally be used only on pre-1949 tractors produced for the domestic agricultural market.

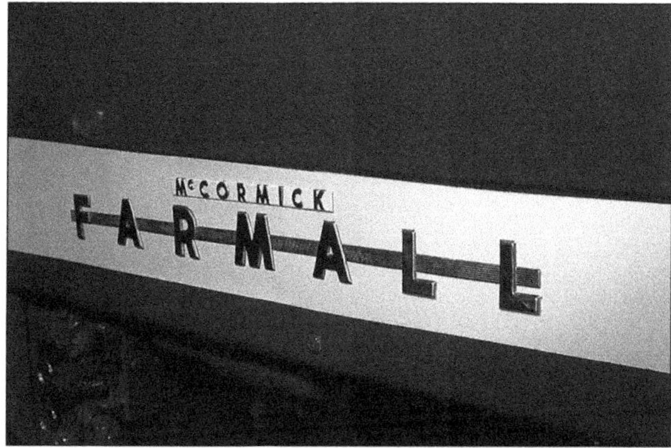

They're not cheap, but there are plenty of sources for reproduction stainless-steel "Farmall" and model number emblems, such as those used on the 400 and 450 models.

Serial Number Plates

For most vintage tractor restorers, restoration of the serial number plate is not only the last step in the project, but also a source of pride. Having a tractor with a low serial number is kind of like acquiring a limited-edition painting with a low number. Consequently, tractor collectors don't take this step lightly.

If you're working with an older-model tractor, you may be lucky enough to have a brass serial number plate. If so, a good polishing with brass cleaner will suffice.

Most of the serial number plates on later-model tractors, though, were made of aluminum. Still, you can make the numbers and letters stand out with a good polishing with steel wool and a quality cleaner. Just be careful on the plates that have a partially black background.

To finish off your restoration, be sure to clean and polish the serial number plate. Aluminum plates can be restored with some cleaner and steel wool. Brass plates, however, are best refinished with a good brass cleaner and polish.

Chapter 18

The Fruits of Your Labor

Although there are a number of tractor restorers who limit their hobby to restoring a desirable model and adding it to their collection, others see restoration as only one step in the pastime. For the former group, collecting antique tractors is not much different than collecting coins or stamps—just a little more expensive. The latter group sees the finished tractor as an avenue to get more involved in the growing number of clubs, tractor shows, and antique tractor events that continue to sprout up in farm country and urban areas alike.

Hopefully, the process up to this point—that of turning a pile of greasy and rusty iron into a showpiece—has been fulfilling on its own. However, there's nothing like the friendships that can be built when you get involved with a group of collectors that share your interests, frustrations, and challenges.

There's nothing quite as fulfilling as that first drive on a tractor you've spent several months restoring. If you don't believe it, just ask Walter Bieri of Savannah, Missouri, who spent most of one winter restoring this Farmall 450.

Antique Tractor Shows

Each year, tractor enthusiasts put together literally hundreds of antique tractor shows throughout the United States, Canada, and Europe. While some of them are sponsored by clubs and cater to a certain brand of tractor, others welcome all tractor brands and are open to both steam- and gasoline-powered models. One of the biggest Farmall shows is the Annual Red Power Round Up, which is held in a different location each year. In the meantime, there are also numerous shows that have a featured brand or tractor each year. That means that someplace, sometime Farmall tractors will be the featured brand at a show near you.

In many cases, you don't even have to take a fully restored model. Go to any one of the shows, and you're sure to see at least one or two tractors setting in the lineup that run just fine, but haven't seen a new coat of paint since they left the factory.

What you will also see are groups of men and women sitting under the closest shade tree or portable canopy, sharing stories, and catching up on each others lives since the last time they all got together. To your benefit, many of them also have tips to share about how they solved a particular problem or located a certain part. You'll find that you're not alone in the challenges you face as show participants share their war stories.

If you're still in the process of restoring a vintage tractor, a tractor show will often give you the opportunity to closely inspect a like model and ask questions of its owner. Assuming that person has restored his tractor to original condition, there's nothing like physically examining the real thing to know what yours is supposed to look like.

If those aren't reasons enough to participate in a tractor show, consider that many shows feature a swap meet or flea market where you can purchase parts for various models of antique tractors. Some of the shows also feature field demonstrations, tractor parades, and tractor games in which you can participate.

Farm Equipment Demonstrations

As the interest in antique tractors has grown, so has the interest in antique farm implements and field demonstrations—and it's easy to see why. When you've got a vintage farm tractor that purrs like a kitten, it's hard to be content just driving it in parades or showing it off to friends. There's always the urge to put it to work in a nostalgic setting.

Hence, many antique tractor shows now feature field demonstrations as part of the agenda. While many started off with plowing demonstrations, the list of activities has grown to include such activities as threshing, baling, shelling corn, cutting silage, and so on. Some, of course, will be stationary demonstrations that use the tractor's belt pulley to power the machine. Others are in the field, where antique tractors can be seen pulling implements that are appropriately matched for the time period and power requirements.

Whether you're watching or participating, it's important to remember one thing: Farm implements built in the early part of this century were not equipped with the safety features, nor the shields and guards, found on today's equipment. Carelessness could cost a finger, arm, or a life at the blink of an eye. So keep your distance from working machines. If you're operating the equipment yourself, or helping a friend, never attempt to make adjustments or clear out a crop slug without first shutting off the tractor. Remember, too, that tractors of the past were never designed for passengers. A tractor fender is not a seat!

Whether it's displayed at the local fair or a national show, a well-restored tractor can be a source of pride for its owner.

Tractor Games

Gather a bunch of antique tractor enthusiasts together and they're bound to come up with other ways to show off their tractors than just a parade of equipment or a series of field demonstrations. Today, it's not unusual to see tractor shows that list such unusual activities as antique tractor square dances, slow races, and backing contests. All are designed to extend the fun associated with owning and restoring antique tractors.

A slow race, for example, tests not only your tractor-operating skills, but your mechanical skills, as well. The goal of the "race" is to see who can drive a certain distance in the slowest time without stopping and without killing the engine. That means you have to decide which gear to start in, how far you dare throttle the engine back, and how often to apply the brakes. The smoother the engine runs at low speeds, the better you're going to be able to challenge the competition.

If maneuvering a tractor is your forte, you can find plenty of tractor games to test your skills in that area, too. The barrel race, for instance, calls for participants to push a barrel, which has been padded to protect tractor finishes, through an obstacle course. Naturally, drivers with narrow-front tractors tend to hold the advantage.

On the other hand, maybe you'd rather try backing. A couple of contest variations include backing a hay wagon up to a pretend loading dock or through an obstacle course. The winner in either case is the person with the fastest time. Another backing skill game requires participants to see who can back up and stop with the hitch positioned closest to an egg without breaking it. This game is designed to test your skills at hitching up an implement and your ability to line up the hitch pin holes. Occasionally, just to level the playing field, the game organizers require all participants to use the same tractor.

There are other games being played by tractor clubs all over the country. The latest, ever growing in popularity, is tractor square dancing. It takes a lot of room, because it is done a lot like real square dancing. A caller announces the moves to the tune of music, and the tractors drive in a circle and follow the calls.

The goal of some restorers is to simply add to their collection, which often encompasses a certain tractor style or series. This assembly of Farmall and McCormick tractors belongs to Walter Bieri.

Tractor shows are a good place to find other models like your own and pick up tips and ideas from other collectors.

The parade of colors and models is a customary part of many tractors shows. Just remember that open platform tractors were designed for only one occupant. Never carry extra passengers on the seat, platform, or, worst of all, the fender.

Due to the growing popularity of antique tractor rides, some tractors have been fitted with an additional seat. Those models with Culti-Vision are ideal candidates for side-by-side seating.

Tractor shows, such as this gathering at the National Red Power Roundup, are an ideal place to meet other collectors and show off your accomplishments.

Antique Tractor Pulls

There are a several reasons antique tractor pulling has become popular with tractor enthusiasts. The first is cost. Compared to modern-day pulling tractors, which are often equipped with multiple turbochargers and high-priced tires, an antique pulling tractor looks pretty much like the original. The only difference is the extra weight racks and the wheelie bars found on some tractors.

Consequently, an antique tractor enthusiast can participate in the sport for a fraction of the cost. On top of that, since the tractors in some classes are not significantly modified, they can still be used for work around the farm or acreage. And most antique tractor pullers simply prefer the slow pace of antique pulls to the glitz, smoke, and noise of modern tractor pulls.

In general, most antique tractor pulls are divided into four or five classes, depending upon the governing body. It used to be that pulling tractors were classified as "antique" if they were built in 1938 or before, while tractors produced from 1939 to 1954 were designated as "classic." The dividing year had more to do with tractor history than anything else. Prior to 1939, things like the radiator, fuel tank, and steering rod were pretty much left exposed, while later-model tractors were more streamlined and typically had more power.

Today, the National Antique Tractor Pullers Association (NATPA) and United States Antique Pullers (USAP), which generate the rules followed by most sanctioned pulls, have taken both history and tractor size into consideration. They realized that fewer and fewer tractors built prior to 1939 were participating in pulls. Hence, both the NATPA and USAP now have five classes.

Division I in the NATPA, for example, is designed for beginning pullers and show tractors and is used to promote stock pulling. Only tractors built in 1959 or earlier, or production models that started in 1959, are eligible. In addition, almost everything must be stock, and tractors can only pull in first gear with a 3-mph speed limit.

Division II is for near-stock tractors that are 1959 or older models. This time, any gear is permitted, and a maximum of 10 percent over stock rpms is allowed. By the time you get into Divisions III and IV, drivers are permitted to use any gear, and any kind of cut is allowed on tires. Division III has a 4-mph speed limit, while IV has an 8-mph limit and allows shifting of a torque amplifier.

Finally, Division V is for non-diesel tractors that are 1961 or older. By this stage of the game, any gear and speeds up to 12 mph are allowed. So is turning up the engine to no more than 30 percent over stock rpm.

No matter what division you participate in, all tractors are required to have a wheelie bar and kill switch. There are different weight classes in each division as well.

Meanwhile, tractors competing in USAP events also have five classes to choose from: Super Farm Stock, Modified Stock, Pro Stock, Super Pro Stock, and Open Pro Stock. Even though there are speed limits in each class, ranging from 3 to 12 mph, all five classes are limited to tractors that are 1959 or older models.

Tractor restoration and tractor shows are often addictive, affecting other members of the family.

Keep in mind that there are dozens of tractor pulling organizations around the country, and many of them operate under their own rules and regulations. The NATPA and USAP continue to review and update regulations as well. It's best to check the rules at any pull before you go to the trouble of loading up the tractor and driving any distance.

Just as with modern tractor pulls, antique tractors in each class pull a mechanical sled, on which the weight increases as the sled is pulled down the track. However, there are basically two ways antique tractor pulls are run. One is by weight class and the other is by percentage pull. Weight classes for antique tractors generally start at 3,500 pounds and increase in 500-pound increments, usually up to 7,500 to 8,500 pounds. Determining the winner of each class is just as simple as it is in big pulls—the tractor pulling the sled the farthest wins.

Percentage pulls, on the other hand, require a little bit of math. In this case, you generally want the tractor to be as light as you can get it, or at least at the lighter end of the class. Consequently, any frills added for cosmetic reasons are often left off. Tractors are then weighed to determine their exact weight and participants attempt to pull as much weight as possible. In the end, if two tractors pull the same amount of weight the same distance, but one weighs 3,500 pounds and the other weighs 4,000 pounds, the lighter tractor would be declared the winner.

The one thing most antique pulls do have in common is the ultimate goal. And surprisingly, it's not the trophy or prize money. What most participants are really after is the pride that comes with winning and the ability to settle the age-old argument—if only for a day—of whether their Farmall can beat their friend's John Deere or Allis-Chalmers.

This custom-built seat, which can be mounted to the three-point hitch, makes it easy for the tractor owner to carry an extra passenger when necessary. There's no word on whether he puts his wife or mother-in-law up there.

Although most antique tractor pulls allow only stock tractors, some include classes for modified tractors, which feature a number of alterations to the original model. Antique tractor pulls, which often divide vintage models into several weight and model year categories, are a popular way to showcase your mechanical and driving skills.

As the interest in antique tractors has grown, so has the interest in antique farm implements. In this case, an old I H potato digger was restored to complement a collector's Model H.

In the 1950s, International sponsored its Farmall Fast-Hitch Square Dance Kids, which toured North America performing their mechanized versions of the tried-and-true hoe-down at state and county fairs everywhere. Moving at a snail-like 3 to 5 mph, the quartet of Farmall 200 tractors "danced" their repertoire of twenty basic square dance "steps." If this didn't sell you on the fleet-footedness and tight turning circle of the Farmall, nothing would.

Richard Porter has made a sideline out of photographing vintage tractors and their owners at tractor shows.

With a little help from their parents, some individuals develop an interest in vintage tractors at a young age.

Bob Thomas took his love for Farmall tractors to the point he built a half-scale version of an F-20, complete with replica engine that runs like the original.

FARMALL

FARMALL
FARMALL

Attachments, including mid- or front-mounted cultivators, listers, and hillers, are also popular with Farmall enthusiasts.

Appendix

Sources for Parts, Rebuilding, and Repairs

PARTS SOURCES

General Parts

Abilene Machine
Box 129
Abilene, KS 67410
Physical address
407 Old Hwy 40, Solomon, KS 67480
877-800-1238
sales@abilenemachine.com
www.abilenemachine.com

A-C Tractor Salvage
8480 225th Avenue
Maquoketa, IA 52060
563-652-2949

Aldermans Inc.
1380 South M-13
4055 South Sheridan Rd.
Lennon, MI 48449
www.aldermans.com

Ag Tractor & Koozer Supply
9301 Breagan Rd.
Lincoln, NE 68526
402-421-8822

All States Ag Parts
(New, used, and rebuilt tractor parts)
Multiple Locations in U.S. and Canada
866-609-1260
www.tractorpartsasap.com

Biewer's Tractor Salvage
16242 140th Ave.
Barnesville, MN 56514
218-493-4696
bts@rrt.net
salvagetractors.com

Bob Martin Antique Tractor Parts
5 Ogle Industrial Drive, Lot #3
Vevay, IN 47043
812-427-2622
antiquetractor@earthlink.net
www.antiquetractorparts.com

Central Michigan Tractor & Parts
2713 N. U.S. 27
St. Johns, MI 48879
800-248-9263

Central Plains Tractor Parts
2116 S Minnesota Ave #4
Sioux Falls, SD 57105
800-234-1968
www.centralplains.com

Colfax Tractor Parts
10447 Field Ave.
Colfax, IA 50054
800-284-3001
info@colfaxtractorparts.com
colfaxtractorauctions.com

Cook Tractor Parts
207 NW 160th
Clinton, MO 64735
800-769-5823
660-885-2287
jrcook@cooktractorparts.com
www.cooktractorparts.com

Cross Creek Tractor Co.
4315 HWY. 278 East
Cullman, AL 35055
800-462-7335
256-739-0496
info@crosscreektractor.com
crosscreektractor.com

Denny's Carb Shop
8620 Casstown-Fletcher Rd.
Fletcher, OH 45326
937-368-2304
dennystractorproducts@yahoo.com
dennyscarbshop.com

Eiklenborg Salvage, Inc
12732 G Ave.
Aplington, IA 50604
319-347-5510
michaeleiklenborg@hotmail.com
eiklenborgsalvage.com

Fort Atkinson Tractor Parts
2645 Highway 24 North
Fort Atkinson, IA 52144
877-530-3010

Gillaspy Tractor Salvage
53789 200th Trl.
Chariton, IA 50049
641-862-3217

Iowa Falls Tractor Parts
Rt. 3, Box 330A
Iowa Falls, IA 50126
800-232-3276

Jensales Tractor Manuals & Parts
200 Main Street
Manchester, MN 56007
800-443-0625
www.jensales.com

Tom Klumpp Equipment & Tractor Parts
4160 Basile Eunice Hwy.
Basile, LA 70515
877-430-4430
337-432-5804
tkequip@centurytel.net
www.tomklumppequip.com

MacFadden and Sons, Inc.
1457 Hwy. Rte. 20
Sharon Springs, NY 13459
518-284-2090
jim@macfaddens.com
www.macfaddens.com

Owosso Tractor Parts
1003 S Washington St.
Owosso, MI 48867
989-729-6567
www.owossotractorparts.com

Pete's Tractor Salvage, Inc.
2163 15th Ave. NE
Anamoose, ND 58710
800-541-7383
701-465-3274
ptractor@gondtc.com
www.petestractor.com

Restoration Supply
96 Mendon St.
Hopedale, MA 01747
800-809-9156
508-634-6915
sales@tractorpart.com
www.tractorpart.com

South-Central Tractor Parts
3806 Old Highway 61 South
Leland, MS 38756
662-686-4131

Southeast Tractor Parts
14720 Highway 151
Jefferson, SC 29718
888-658-7171
843-658-7171
setractor@shtc.net
partslady@shtc.net
setractorparts.com

Steiner Tractor Parts, Inc.
1660 S. M-13
Lennon, MI 48449
800-234-3280
www.steinertractor.com

TractorSpot
www.tractorspot.com

Van Noort Tractor Salvage
1003 10th Ave.
Rock Valley, IA 51247
800-831-8543

Walt's Tractors
2634 Audrain Road 381
Mexico, MO 65265
888-414-4043
www.waltstractors.com

Wengers of Myerstown
814 South College St.
Myerstown, PA 17067
800-451-5240
717-866-2135
www.wengers.com

Wilson Farms
20552 Old Mansfield Rd.
Bellville, OH 44813
740-694-5071

Worthington Tractor Salvage
Headquarters
5300 N. Annika Ave.
Sioux Falls, SD 57107
Additional locations in WI, IA, MI, MO, SD, IN, MN & NC
888-845-8456
inetcomments@worthingtonagparts.com
www.worthingtonagparts.com

Yesterday's Tractors
751 Commerce Loop
Port Townsend, WA 98368
800-853-2651
www.yesterdaystractors.com

Specialized Parts

The Brillman Company
2328 Pepper Road
Mt. Jackson, VA 22842
888-274-5562
540-477-4112
www.brillman.com
Wiring harnesses, electrical components, etc.

Diesel Power & Machine
7 Matchett Drive
Pierceton, IN 46562
800-825-7711
574-594-5888
info@dieselpower-reman.com
www.dieselpower-reman.com
Remanufactured diesel engines

The Fordson House
717 Stephenson Avenue
Escanaba, MI 49829
906-786-5120
info@thefordsonhouse.com
www.thefordsonhouse.com
Fordson, International, Farmall, McCormick-Deering, and IHC antique tractor parts

Franzen Family Tractor Parts
1218 49th St.
Monmouth, IA 52309
563-673-6631
IH parts and restoration

J.P. Tractor Salvage
1347 Madison 426
Fredericktown, MO 63645
573-783-7055
jpparts@jptractorsalvage.com
www.jptractorsalvage.com
International Harvester and Farmall parts

COMPONENT PARTS/REBUILDERS

Sleeves and Pistons

Carr's Repair
Box 219
Barwick, Ontario, Canada
P0W 1A0
Physical address
1801 2nd Ave. PMB 497
International Falls, MN 56649
807-487-2548
gandjcarr@tbaytel.net
www.carrsrepairvintageparts.com
IH and John Deere sleeves & pistons

Carburetors and Governors

Denny's Carb Shop
8620 Casstown-Fletcher Rd.
Fletcher, OH 45326
937-368-2304
dennystractorproducts@yahoo.com
www.dennyscarbshop.com

Link's Carburetor Repair
P.O. Box 181
8708 Floyd Hwy. North
Copper Hill, VA 24079
540-929-4519

McDonald Carb & Ignition
1001 Commerce Rd.
Jefferson, GA 30549
706-367-4179
info@mcdonaldcarb.com
www.mcdonaldcarb.com
Electrical and fuel system parts

Motec Engineering
7342 IN-28
Tipton, IN 46072
765-963-6628

Robert's Carburetor Repair
1601 35th Ave. W
Spencer, IA 51301
712-262-5311
info@robertscarbrepair.com
www.robertscarbrepair.com
Carburetor parts and repair

Treadwell Carburetor Company
4870 County Highway 14
Treadwell, NY 13846
607-829-8321
carbkit@frontiernet.net
www.carbsandkits.com

Diesel Injection Pumps and Nozzles—Parts/Rebuilders

Central Fuel Injection Service Co.
2403 Murray Rd.
Estherville, IA 51334
712-362-4200
service@centralfuel.com
www.centralfuelinjectionservice.com

Guenther Heritage Diesel
313 Red Oak Rd.
Walnut, IL 61376
815-303-3492
ed64drag@gmail.com

Magnetos—Parts/Rebuilders

Kevin's Magneto Service
25930 N. County Road, 2600 E
Manito, IL 61546
309-303-2634
Kevinsmags@yahoo.com
www.magneto-repair.com

Mark's Carburetor & Magneto Services
2189 125th St. NW
P.O. Box 15
Rice, MN 56367
320-393-7272
www.marks-carburetor-service.square.site

Lightning Magneto
Mitch Malcolm
45685 Co. Hwy 54
Ottertail, MN 56571
218-367-2819
mitch@lightningmagneto.com
www.lightningmagneto.com

Fasteners

Bolt Depot
100 Research Road
Hingham, MA 02043
(order pickups only)
866-337-9888
info@boltdepot.com
www.boltdepot.com

Cab Interiors

Fehr Cab Interiors
10116 N 1900 E Rd
Fairbury, IL 61739
815-692-3355
sales@fehrcab.com
www.fehrcab.com

Tractor Interior Upholstery LLC
4979 Orchid Ave.
Northwood, IA 50459
641-390-0321
www.tractorinteriors.com

Wiring Harnesses

Agri-Services
13899 North Rd.
Alden, NY 14004
716-937-6618
www.wiringharnesses.com

Seals and Gaskets

Lubbock Gasket & Supply
402 19th St.
Lubbock, TX 79401
800-527-2064
806-763-2801
www.lubbockgasket.com

Olson's Gaskets
3059 Opdal Road E.
Port Orchard, WA 98366
360-871-1207
info@olsongaskets.com
www.olsonsgaskets.com

Wheels and Rims

Nielsen Spoke Wheel Repair
Herb Nielsen
3921 230th St.
Estherville, IA 51334
712-867-4796

TNT Poly Div.
Taube Tool Corp.
1524 Chester Blvd.
Richmond, IN 47374
765-962-7415
Replacement lugs for steel wheels

Tires

M. E. Miller Tire Co.
17386 State Hwy. 2
Wauseon, OH 43567
800-621-1955
419-335-7010
www.millertire.com

Tucker's Tire Company
844 S. Main St.
Dyersburg, TN 38024
888-248-7146
731-285-8520
tires4u@webtv.net
www.tuckertire.com

Replacement Seats/ Cab Interior

K&M Manufacturing
P. O. Box 409
308 NW 2nd St.
Renville, MN 56284
800-328-1752
www.tractorseats.com

Decals

K & K Antique Tractors
5995 N. 100 W.
Shelbyville, IN 46176
317-398-9883
www.kkantiquetractors.com

Maple Hunter Decals
P. O. Box 805
Riley, IN 47871
812-894-9759
maplehunterdecals@gmail.com
www.maplehunterdecalsindiana.com
Decals for a wide variety of applications and models

Restoration Equipment

CJ Spray, Inc.
2845 W. Service Rd.
Eagan, MN 55121
888-257-7729
customerservice@cjspray.com
www.cjspray.com
Airless Paint Spray systems

K & S Steel Products
4620 S Co Rd 550 E
Greensburg, IN 47240
812-663-8567
Engine & tractor splitting stands

TP Tools and Equipment
Tip Plus Corp.
7075 State Route 446
Canfield, OH 44406
800-321-9260
Tech Line: 330-533-3384, Ext. 22
www.tptools.com
(Parts washers, grinders, presses, sandblasting equipment, etc.)

PUBLICATIONS

Tractor Manuals

Jensales Inc.
200 Main Street
Manchester, MN 56007
800-443-0625
www.jensales.com

Goodburn Literature Sales
79848 290th Rd.
Madelia, MN 56062
507-642-8481

Yesterday's Tractors
751 Commerce Loop
Port Townsend, WA 98368
800-853-2651
www.yesterdaystractors.com

The Manual Store
323 Prairie View Dr.
Carroll, IA 51401
sales@themanualstore.com
www.themanualstore.com

Abilene Machine
Box 129
Abilene, KS 67410
(Physical Address)
407 Old Hwy 40, Solomon, KS 67480
877-800-1238
sales@abilenemachine.com
www.abilenemachine.com
I & T Shop Manuals

General Magazines

Ageless Iron Almanac
Meredith Corp.
1716 Locust Drive
Des Moines, IA 50309
515-284-3000
saicustser@cdsfullfillment.com
www.agelessiron.com/myaccount

Antique Power
Circulation
P. O. Box 292026
Kettering, OH 45429
888-760-8108
antiquepower@sfsdayton.com
www.antiquepower.com

Heritage Iron
P. O. Box 519
Greenville, IL 62246
855-653-4766
info@heritageiron.com
www.heritageiron.com

The Hook Magazine
P. O. Box 2373
Elizabethtown, KY 42701
270-202-6742
www.hookmagazine.com
Tractor pulling, including antique and classic

Vintage Tractor Digest
P. O. Box 10
Bethlehem, MD 21609
410-673-2414
stant@threshermen.org
www.vintagetractordigest.com

Clubs and Brand Newsletters/ Magazines

INTERNATIONAL HARVESTER
National International Harvester
Collectors Club, Inc.
Membership Department
P.O. Box 35
Dublin, IN 47335
765-478-6179
ihcclub@aol.com
www.nationalihcollectors.com

Red Power Magazine
P.O. Box 245
Ida Grove, IA 51445
712-364-2131
dmiesner@netllc.net
www.redpowermagazine.com

A pre-production 1932 Farmall F-12. (Photograph by Ralph W. Sanders/Motorbooks Archive)

A nicely restored Farmall 340 high-crop machine. (Photograph by Chester Peterson Jr.)

Index

About the Author

Tharran E. Gaines was born in north-central Kansas where he grew up on a small grain and livestock farm near the town of Kensington. The only boy in a family of five children, he often claims that he did all of the farm work normally delegated to a full farm family of boys. Others say he was spoiled by four younger sisters.

He attended Kansas State University, where he received a degree in wildlife conservation and journalism with the goal of going into outdoor writing. He soon was using his agricultural background as a technical writer for Hesston Corporation and hasn't left agriculture since.

As a technical writer, he has produced repair manuals, owner's manuals, and assembly instructions for Hesston, Winnebago, Sundstrand hydrostatic transmissions, Kinze planters and grain wagons, and Best Way crop sprayers. As a creative writer, he has crafted and produced everything from newsletter and feature articles to radio and TV commercials to video scripts and advertising copy for such companies as DeKalb Seed Company, DowElanco, Asgrow Seed Company, Farmland Industries, Case IH, Caterpillar, AGCO, Gleaner, Challenger and Massey Ferguson.

In 1991, he began his own business as a freelance writer, and today he continues to operate Gaines Communications with his wife Barb out of their home office. The majority of their business since that time has involved producing editorial copy for AGCO Corporation's company magazines, including the AGCO Allis *Landhandler* and Massey Ferguson *FarmLife* publication. He is also serves as a contributing writer for *Successful Farming*, *Ageless Iron*, and *Heritage Iron* magazines.

Tharran and his wife live in a 100-plus-year-old Victorian home in Savannah, Missouri.

www.ingramcontent.com/pod-product-compliance
Lightning Source LLC
LaVergne TN
LVHW060638110826
845147LV00018B/1003

* 9 7 8 0 7 6 0 3 6 8 9 6 1 *